Collins

The Shanghai Maths Project

For the English National Curriculum

Series Editor: Professor Lianghuo Fan
UK Curriculum Consultant: Jo-Anne Lees

Practice Book

William Collins' dream of knowledge for all began with the publication of his first book in 1819. A self-educated mill worker, he not only enriched millions of lives, but also founded a flourishing publishing house. Today, staying true to this spirit, Collins books are packed with inspiration, innovation and practical expertise. They place you at the centre of a world of possibility and give you exactly what you need to explore it.

Collins. Freedom to teach.

Published by Collins
An imprint of HarperCollins*Publishers*
The News Building, 1 London Bridge Street, London, SE1 9GF, UK

HarperCollins*Publishers*
Macken House, 39/40 Mayor Street Upper, Dublin 1, D01 C9W8, Ireland

Browse the complete Collins catalogue at collins.co.uk

10 9 8 7 6 5 4 3 2

ISBN 978-0-00-814472-2

The Shanghai Maths Project (for the English National Curriculum) is a collaborative effort between HarperCollins, East China Normal University Press Ltd. and Professor Lianghuo Fan and his team. Based on the latest edition of the award-winning series of learning resource books, *One Lesson One Exercise*, by East China Normal University Press Ltd. in Chinese, the series of Practice Books is published by HarperCollins after adaptation following the English National Curriculum.

Practice book Year 11 is translated and developed by Professor Lianghuo Fan with assistance of Ellen Chen, Haiyang Ding, Kunli Li, Dr. Na Li, Jietong Luo, Ming Ni, Nianbing Ren, Qi Tang, Wanguo Tian, Dr. Yi Wang, Sicheng Xie, Jing Xu, Daolu Yao and Zaiyang Yu, with Jo-Anne Lees as UK curriculum consultant.

British Library Cataloguing-in-Publication Data
A catalogue record for this publication is available from the British Library.

Series Editor: Professor Lianghuo Fan
UK Curriculum Consultant: Jo-Anne Lees
Publisher: Katie Sergeant
Content Editor: Daniela Mora Chavarria
Copyeditor: Julie Bond
Proofreader: Catherine Dakin
Answer checker: Mike Fawcett and Steven Matchett
Cover designer: Steve Evans
Typesetter: East China Normal University Press Ltd.
Production controller: Alhady Ali
Printed and Bound in the UK using 100% Renewable Electricity at CPI Group (UK) Ltd

For more information visit:
www. harpercollins. co. uk/green
Acknowledgements

The publishers gratefully acknowledge the permission granted to reproduce the copyright material in this book. Every effort has been made to trace copyright holders and to obtain their permission for the use of copyright material. The publishers will gladly receive any information enabling them to rectify any error or omission at the first opportunity.

MIX
Paper from responsible sources
FSC™ C007454

This book is produced from independently certified FSC™ paper to ensure responsible forest management.

Contents

Chapter 1 Real numbers

1.1 Calculation with real numbers: Revision

Learning objective

Calculate with roots, and with integer and fractional indices; calculate with numbers in standard form.

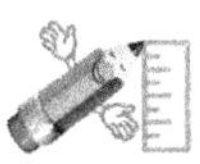

A. Multiple choice questions

1 How many times as great as 3^{12} is 9^6?

A. 1

B. $\left(\frac{1}{3}\right)^2$

C. $\left(\frac{1}{3}\right)^6$

D. $(-6)^2$

2 Calculate the value of $\frac{1}{2} + \frac{2}{3} + \frac{3}{4} \times (-4)$.

A. -1

B. $-\frac{11}{6}$

C. $-\frac{12}{5}$

D. $-\frac{23}{3}$

3 The topic for a mathematics lesson is Pythagoras' theorem. Antoni typed Pythagoras' theorem into a search engine and it displayed about 12 500 000 results. This number can be expressed in standard form as (   ).

A. 1.25×10^5

B. 1.25×10^6

C. 1.25×10^7

D. 1.25×10^8

4 How many of these numbers are irrational?

π $\qquad$ $\frac{1}{3}$ $\qquad$ $\sqrt{2}$ $\qquad$ $\sin 30°$

A. one $\qquad$ B. two $\qquad$ C. three $\qquad$ D. all

5 The diagram shows rectangle $OABC$, drawn on a number line. The length of OA is 2 units and the length of AB is 1 unit. An arc is drawn with O as the centre and OB as the radius. What is the exact value of the point of intersection of the arc with the number line?

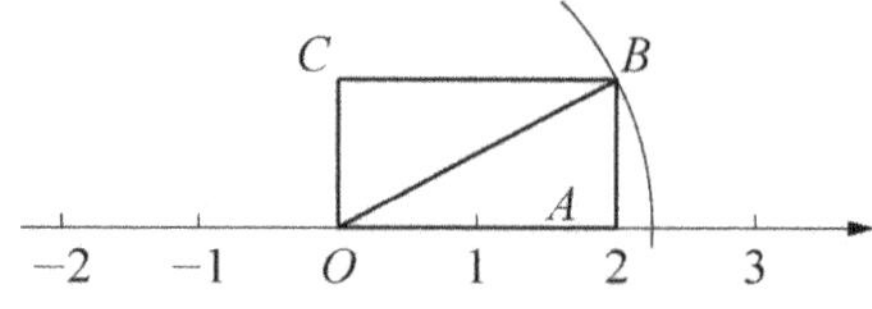

A. 2.5 B. $2\sqrt{2}$ C. $\sqrt{3}$ D. $\sqrt{5}$

B. Fill in the blanks

6 According to the Richter Scale, the relationship between the seismic energy E that is released by an earthquake and its magnitude on the scale n is $E = 10^n$. Therefore, the seismic energy released by a 9.0-magnitude earthquake is ________ times the seismic energy released by a 7.0-magnitude earthquake.

7 Look at the following pattern:

$$\frac{1}{1} + \frac{1}{2} - 1 = \frac{1}{2}$$

$$\frac{1}{3} + \frac{1}{4} - \frac{1}{2} = \frac{1}{12}$$

$$\frac{1}{5} + \frac{1}{6} - \frac{1}{3} = \frac{1}{30}$$

$$\frac{1}{7} + \frac{1}{8} - \frac{1}{4} = \frac{1}{56}$$

Complete this stage of the pattern: $\frac{1}{2011} + \frac{1}{2012} -$ ________ $= \frac{1}{2011 \times 2012}$.

8 x and y are real numbers, that satisfy $\sqrt{1+x} - (y-1)\sqrt{1-y} = 0$.
Therefore $x^{2011} - y^{2011} =$ ________.

9 Calculate, simplifying any surds: $\left(-\frac{1}{2}\right)^{-1} + (\pi - \sqrt{2019})^0 + \sqrt{12} =$ ________.

10 Calculate, simplifying any surds: $(-3)^0 - \sqrt{27} - 1 - \sqrt{2} - \dfrac{1}{\sqrt{3}+\sqrt{2}} =$ ________.

11 Calculate: $\left\{\left[3\dfrac{3}{4} \div \left(\dfrac{3}{4} - 1\right) + (1 - 0.6) \times \left(\dfrac{2}{5}\right)^2\right] \div \left(-\dfrac{5}{3}\right) - 20\right\} \times (-1)^{39} =$ ________.

C. Questions that require solutions

12 Look at this pattern:

$\dfrac{1}{1\times 2} = 1 - \dfrac{1}{2}$; $\dfrac{1}{2\times 3} = \dfrac{1}{2} - \dfrac{1}{3}$; $\dfrac{1}{3\times 4} = \dfrac{1}{3} - \dfrac{1}{4}$; ......

Then answer the following questions.

(a) If n is a positive integer, then make a hypothesis: $\dfrac{1}{n(n+1)} =$ ________.

(b) Prove the conclusion obtained from your hypothesis.

(c) Find the sum: $\dfrac{1}{1\times 2} + \dfrac{1}{2\times 3} + \dfrac{1}{3\times 4} + \cdots + \dfrac{1}{2009\times 2010}$.

1.2 Converting between recurring decimals and fractions

Learning objective

Work interchangeably with terminating decimals and their corresponding fractions.

A. Multiple choice questions

1 Which of the following statements is correct? (   )

A. Some recurring decimals cannot be expressed as fractions.

B. All recurring decimals can be expressed as fractions.

C. All decimals can be expressed as fractions.

D. No fractions can be converted to non-recurring decimals.

2 The sum of two recurring decimals can be expressed as (   ).

A. a whole number

B. a fraction

C. a recurring decimal

D. a whole number, a fraction or a recurring decimal

B. Fill in the blanks

3 Convert the following fractions into decimals.

(a) $\frac{2}{5}$ = ________ (b) $\frac{5}{7}$ = ________

(c) $\frac{4}{3}$ = ________ (d) $\frac{30}{9}$ = ________

4 Convert the following recurring decimals to fractions.

(a) $0.\dot{1}$ = ________ (b) $0.\dot{2}\dot{3}$ = ________

(c) $0.\dot{8}\dot{9}$ = ________ (d) $2.\dot{3}$ = ________

5 Calculate, giving your answer as a fraction: $0.\dot{4} + 1.\dot{2}\dot{3}$ = ________.

6 Calculate, giving your answer as a recurring decimal: $0.\dot{7}8\dot{6} - 0.786$ = ________.

C. Questions that require solutions

7 Calculate: $3.\dot{0}0\dot{8} + 5.05\dot{3}\dot{4} + 19\frac{991}{999} + 8\frac{9371}{9900} =$ ________.

8 Write two different recurring decimals, so that:

(a) their sum is a whole number: ____________________

(b) their product is a whole number: ____________________

(c) their quotient is a whole number: ____________________

1.3 Rounding and truncating

Learning objective

Round numbers and measure to an appropriate degree of accuracy; apply and interpret limits of accuracy, including upper and lower bounds.

A. Multiple choice questions

1 Look at the following numbers. After ( ) is rounded to 2 decimal places, the result is 20.01.

A. 20.004 B. 20.015 C. 20.001 D. 20.009

2 When 0.027 04 is rounded to 3 significant figures, the result is ( ).

A. 0.02 B. 0.03 C. 0.0270 D. 0.027

3 Look at the following numbers. After ( ) is truncated to 2 decimal places, the result is 15.23.

A. 15.2301 B. 15.2259
C. 15.2003 D. none of these numbers

4 When 10 332 is truncated to 3 significant figures, the result is ( ).

A. 10 000 B. 10 300 C. 10 330 D. 10 332

5 Compare these two approximate values: 5.1 kg and 5.10 kg. Which of the following statements is correct? ( )

A. They have the same degree of accuracy.
B. The lower bounds of the two values are equal.
C. The upper bounds of the two values are equal.
D. None of the above is correct.

B. Fill in the blanks

6 Round each of the following numbers to the given degree of accuracy.

(a) 0.006 730 ≈ ________ (to three significant figures)
(b) 12.581 ≈ ________ (to the nearest hundredths place)

(c) 5 364 789 ≈ ________ (to two significant figures)

(d) 92 343 ≈ ________ (to the nearest hundred)

7 Complete the following statements.

19.953 rounded to two significant figures is ________.

19.953 truncated to two significant figures is ________.

19.953 rounded to 1 decimal place is ________.

19.953 truncated to 1 decimal place is ________.

8 2340 kilograms = ________ tonnes ≈ ________ tonnes (truncated to the nearest 1 tonne).

9 3 litres 130 millilitres = ________ litres ≈ ________ litres (truncated to the nearest 1 litre).

10 The lower bound of 8.5 cm, measured to the nearest 0.1 cm, is ________ cm, and the upper bound is ________ cm.

11 60 pieces of paper are used to make a notebook. The greatest number of notebooks that can be made with 2800 pieces of paper is ________.

C. Questions that require solutions

12 A child's outfit uses 2.2 metres of fabric. How many of these outfits can be made with 50 metres of fabric?

13 A box can hold 4.5 kilograms of rice. How many boxes are needed to hold 62 kilograms of rice?

14 Kira has a mass of 70 kg (rounded to the nearest 10 kg) and her younger brother Koji has a mass of 48 kg (rounded to the nearest 1 kg). What is the smallest possible difference in their masses? What is the largest possible difference?

1.4 Solving ratio problems with fractions

Learning objective

Identify and work with fractions in ratio problems; solve problems involving compound measures such as speed, scale diagrams and maps.

A. Multiple choice questions

1 Given that salt makes up $\frac{1}{10}$ of salt water, the ratio of salt to water is (　　).

A. 1 : 8　　B. 1 : 9　　C. 1 : 10　　D. 1 : 11

2 Ali has read 60% of the total number of pages of a book. The ratio of the number of pages he has not read to the total number of pages is (　　).

A. 2 : 3　　B. 3 : 5　　C. 2 : 5　　D. 5 : 8

B. Fill in the blanks

3 If Number A is 25% greater than Number B, then the ratio of Number A to Number B is ________.

4 32 kilograms of flour is milled from 40 kilograms of wheat. Based on the same ratio, the number of kilograms of flour that can be milled from 7 tonnes of wheat is ________.

5 A tree produces a shadow on the ground that has a length of 8.4 metres. At the same time, a 2-metre-long vertical pole produces a shadow with a length of 1.2 metres. Therefore the height of the tree is ________ metres.

6 A box contains different coloured pens. $\frac{5}{12}$ of the pens are black. Another 18 black pens are put into the box, and the ratio of the number of black pens to the number of pens with other colours is now 2 : 1. The total number of black pens in the box now is ________.

7 The ratio of the number of people in a choir to the number of people in a dance group is 3 : 2. 10 people move from the choir to the dance group, and the ratio of the people in the choir to people in the dance group is now 7 : 8. The number of people in the choir originally was ________.

C. Questions that require solutions

8 There are 538 workers in a factory that has three workshops. There are 12 fewer workers in the first workshop than in the third workshop. The ratio of workers in the second workshop to the number of workers in the third workshop is 3 : 4. How many workers are there in each workshop?

9 Four families live in a block of flats. The water bill for all the residents in the building for the last month is £392. The number of people in the four families are 2, 4, 3, and 5. If the water bill is shared equally by each resident, how much does each family have to pay for the water bill?

10 The distance between towns A and B is 12 centimetres on a map. The actual distance between these two towns is 480 kilometres. If the distance between city M and city N on the map is 4 centimetres, find the actual distance between the two cities, M and N.

11 A map has a scale of 1 : 6 000 000. On this map, the distance between two towns is 5 centimetres. Two cars, A and B, start travelling at the same time from each of these towns and move towards each other. They pass each other after 3 hours. Given that the ratio of the speeds of the two cars is 2 : 3, find the speed of each car in km/h.

1.5 Solving rate problems with compound units

Learning objective

Solve problems involving, and convert between, compound units such as speed and density in various contexts.

A. Multiple choice questions

1 The unit for measuring ( ) is a compound unit.

A. length B. weight C. area D. speed

2 Which of the following statements is correct about the units of (1) volume, (2) wage, (3) density and (4) price? ( )

A. All the units are compound units.

B. The units for (1), (2) and (3) are compound units.

C. The units for (2), (3) and (4) are compound units.

D. The units for (1), (3) and (4) are compound units.

B. Fill in the blanks

3 Convert these units of speed: 3 m/s = ________ km/h, 72 km/h = ________ m/s.

4 Convert these units of density: 1.5×10^3 kg/m^3 = ________ g/cm^3, 1 g/cm^3 = ________ kg/m^3.

5 In a sports competition, a bullet was released, travelling at a speed of 680 m/s, from a pistol which was 17 metres away from a spectator. If the speed of sound in air is 340 m/s, then when the spectator heard the sound of the gun firing, the bullet had travelled ________ metres.

6 Two particles, A and B, are travelling in a straight line at a constant speed. The ratio of the speeds of A and B is 2 : 3 and the ratio of the distances that A and B have covered is 3 : 4. Therefore the ratio of time that A travels to the time that B travels to cover these distances is ________.

7 A 20-metre-long train passes over a 980-metre-long iron bridge at a constant speed of 36 km/h. It takes the train ________ minutes to pass over the bridge.

C. Questions that require solutions

8 A bus started from station A travelling at a speed of 90 km/h and arrived at station B in 20 minutes. Then it travelled at 60 km/h for another 10 minutes and arrived at station C.

(a) What was the distance between the two stations A and C?

(b) What was the average speed at which the bus travelled from station A to station C?

9 The smallest solid substance with the least density, "Areogel", is an important result of scientific research on new materials. The hardness and durability of the material is as good as steel, while it can withstand temperatures as high as 1400 ℃. It has a density of 3 kg/m^3. To build a jumbo jet aircraft currently requires 130 tonnes of steel with a super high density of 7.8×10^3 kg/m^3. What mass of "Areogel" is needed to replace steel to build a jumbo jet?

10 After an empty bottle is filled with water, its total mass is 64 g. If the bottle is emptied and then filled with ethanol, the total mass is 56 g. Find the mass of the empty bottle. (density of ethanol = 800 kg/m^3; density of water = 1000 kg/m^3)

1.6 Solving growth and decay problems

Learning objective

Solve problems and interpret the answers in growth and decay problems, including compound interest.

A. Multiple choice questions

1 A factory increases its production at an average rate of P items a month. The annual growth rate of the factory is (   ).

A. P B. $12P$ C. $(1+P)^{12}$ D. not sure

2 A bike-sharing scheme provides low-cost transport for people in cities. One bike-sharing company makes 1000 bikes available in the first month, and in the third month, it plans to provide 440 more bikes than in the first month. Let the growth rate of the number of bikes the company provides in the second and third months be x. Then the correct equation for calculating the total number of bikes available in the third month is (   ).

A. $1000(1+x)^2 = 1000 + 440$ B. $1000(1+x)^2 = 440$

C. $440(1+x)^2 = 1000$ D. $1000(1+2x) = 1000 + 440$

B. Fill in the blanks

3 A company produced a new type of electronic equipment. The company plans to reduce the cost of the equipment by 36% in two years. The average annual percentage of reduction of the cost will be ________.

4 A machine normally costs £96. After being reduced in price twice, the machine costs £54 in a sale. Each time it is reduced, the average percentage reduction in the price of the machine is ________%.

C. Questions that require solutions

5 A company deposits £800 000 in a bank account. The interest rate of the account is 3% compound interest, which is paid annually.

(a) Calculate the total amount of money the company has in the account after 8 years.
(b) Calculate the interest the company earns in 8 years.

6 A company invests £200 000 per year for 5 years in a construction project. All of the money invested in the project is borrowed from a bank, which charges interest at an annual rate of 6%. Calculate the total amount of the cost of the project, including the interest paid on the loan.

7 An organisation deposits £120000 in a bank account at the beginning of each year for 5 consecutive years. The annual interest rate paid by the bank is 5% compound interest, which is added at the end of each year. Calculate the total amount of money in the account and the total amount of interest earned at the end of the fifth year.

8 Joshua is buying an apartment. He will pay a deposit of £500 000, and then pay the remaining cost using a loan that requires him to repay £50000 at the end of each year, plus interest at an annual rate of 6% for 20 consecutive years. Calculate the total cost of the apartment.

9 The initial mass of a radioactive element is 500 g, and its mass reduces at a rate of 10% each year.
(a) Find the expression of the mass m of the radioactive element after t years.
(b) Use the expression that you wrote in part (a) to find the mass of the element after 1 year, 2 years, and so on, up to 10 years. Write your answers in a table.
(c) Use your table from part (b) to estimate the time taken for the mass of the radioactive element to reduce to half of its initial value.

1.7 Solving equations numerically using iteration

Learning objective

Find approximate solutions to equations numerically using iteration.

A. Multiple choice questions

1 The positive root of the equation $5x^2 - 7x - 1 = 0$ is in the interval (　　).

A. (0, 1)　　B. (1, 2)　　C. (2, 3)　　D. (3, 4)

2 Let x_0 be the solution of the equation $3^x = 5 - x$. Then x_0 is in the interval (　　).

A. (3, 4)　　B. (2, 3)　　C. (1, 2)　　D. (0, 1)

3 Which of the following equations is not a rearrangement of $x^2 + 4x = 9$? (　　)

A. $x = \sqrt{9 - 4x}$　　B. $x = 3 - 2\sqrt{x}$　　C. $x = \frac{9 - x^2}{4}$　　D. $x = \frac{9}{x} - 4$

B. Fill in the blanks

4 An iterative sequence is given by $x_{n+1} = 2x_n - 3$. The first five terms of the sequence are 4, ________, ________, ________, and ________.

5 The population of a city is 88 000 and it is expected to grow by 5% each year for the next 10 years. To work out the population of the city for each of the next 10 years, Sam uses the iterative formula $x_{n+1} = (1 + 5\%)x_n$. The value of $x_0 =$ ________, and the population in five years' time is ________.

C. Questions that require solutions

6 Show that the equation $x^5 - 4x^2 - 1 = 0$ has a solution between $x = 1$ and $x = 2$.

7 Show that the equation $x^3 + 2x = -10$ has a solution between $x = -1.9$ and $x = -1.8$.

8 Show that the equation $3x^3 + 2x - 1 = 0$ can be rearranged to give $x = \sqrt[3]{\frac{1 - 2x}{3}}$

9 Find an approximate solution to the equation $x^3 + 2x^2 + 10x - 20 = 0$ in the interval $\left(1, \frac{3}{2}\right)$, giving your answer to the nearest 0.001.

10 Find a solution to the equation $x = \cos x$ in the interval $(0, 1)$, giving your answer to the nearest 0.01.

11 Find an approximate solution to the equation $2^x + x = 4$, giving your answer to the nearest 0.1.

12 Given the equation $x^3 - 5x + 1 = 0$, answer the following questions.

(a) Show that the equation has a root between $x = 2$ and $x = 2.2$.

(b) Show that the equation can be arranged to give $x = \sqrt[3]{5x - 1}$.

(c) Use the iteration $x_{n+1} = \sqrt[3]{5x_n - 1}$, with $x_0 = 2.2$ to find a solution to the equation, correct to 1 decimal place.

Unit test 1

A. Multiple choice questions

1 How many of the numbers $\sqrt{3}$, $\frac{22}{7}$, $\frac{\pi}{2}$ and $\cos 60°$ are irrational? (　　)

A. one　　B. two　　C. three　　D. all

2 Calculate: $0.\dot{3} + \frac{1}{2} =$ (　　).

A. $\frac{2}{3}$　　B. $\frac{3}{4}$　　C. $\frac{5}{6}$　　D. $\frac{3}{5}$

3 0.2026 rounded to 3 significant figures is (　　).

A. 0.202　　B. 0.203　　C. 0.200　　D. 0.210

4 The capacity of a water bottle is 2.5 litres. How many of these bottles are needed to hold 32.5 litres of water? (　　)

A. 15　　B. 16　　C. 13　　D. 12

5 Nadiya needs three days to make an item to sell at a craft market. She completes one third of the work on the first day and half of the remaining work on the second day. Which of the following statements is correct? (　　)

A. She still has one third of the work to do.

B. She still has half of the work to do.

C. She still has one-sixth of the work to do.

D. You cannot find out how much of her work is remaining.

6 Let x_0 be the solution of the equation $x^3 + 2x = 10$. Then x_0 is in the interval (　　).

A. $(0, 1)$　　B. $(1, 2)$　　C. $(2, 3)$　　D. $(3, 4)$

7 Which of the following equations is not a rearrangement of $4x^2 + 3x = 1$? (　　)

A. $x = \frac{1}{2}\sqrt{1 - 3x}$　　B. $x = \frac{1}{3 + 4x}$　　C. $x = \frac{1}{3} - 4x^2$　　D. $x = \frac{1 - 4x^2}{3}$

8 A shop reduces the price of a camera twice using the same discount rate. The final sale price is £149. The original price before the reductions was £300. Let the discount rate be x. Which of the following equations will calculate the final sale price? (　　)

A. $300(1-x)^2=149$　　B. $300(1-2x)=149$

C. $300(1-x^2)=149$　　D. $300(1+x)^2=149$

B. Fill in the blanks

9 The height of Mount Everest is about 8850 metres. The number can be expressed in standard form as ________.

10 The integers greater than $\sqrt{10}$ and less than $\sqrt{26}$ are ________.

11 Calculate: $(-2)^{-1}-3\cos 30°+(\pi-\sqrt{2022})^0+\sqrt{12}=$ ________.

12 a is an irrational number, $-1<a<2$. Write two possible values of a: ________.

13 Lily is 163 cm tall and her sister Lisa is 142 cm tall, both rounded to the nearest 1 cm. The smallest possible difference in their heights is ________ and the largest is ________.

14 The solution to the equation $x^3-x-2=0$ in the interval $(1, 2)$ is ________ (to the nearest 0.01).

15 Given that x and y are real numbers, and they satisfy $\sqrt{x+1}-\sqrt{(y-4)^2}=0$, then $x^{2022}+y^2=$ ________.

16 Two cuboid-shaped containers have the same capacity. The ratio of their lengths is 2 : 1, and the ratio of their widths is 2 : 3. The ratio of their heights is ________.

17 A ladder of length 2.8 metres leans against a vertical wall. The bottom of the ladder is 0.5 metres from the base of the wall. If the top of the ladder slides 0.3 metres down the wall, then the bottom of the ladder is ________ metres from the base of the wall (to the nearest 0.1).

C. Questions that require solutions

18 Mr and Mrs Smith and their three children went on a boat trip. The cost for each adult was £15 and for each child was £10. The boat travelled 8 km for 90 minutes.

(a) Find the total cost of their tickets.

(b) Find the average speed of the boat in km/h during the first 90 minutes.

(c) The family spent two hours on the boat. What was the distance travelled? (Assume that the boat travelled at the speed you found in part (b).)

19 Andrea travels by bus from her home to visit her grandpa. She leaves home at 3 p. m. The distance between her home and her Grandpa's is 90 kilometres. The bus travels 10 km in the first ten minutes. If the bus continues to travel at this speed:

(a) how long does it take Andrea to reach her Grandpa's house?

(b) at what time does she arrive?

20 A gardening team planted some trees in three days. On the first day, they planted 12.5% of the total number of trees, and on the second day, they planted 120 trees. The ratio of the number of trees that remained to be planted to the number of trees planted by the end of the second day was 3 : 5.

(a) How many trees did the team plant on day three?

(b) What is the percentage of the total number of trees were planted in the first two days?

(c) How many trees were planted altogether?

21 Look at the following pattern:

$$\frac{1}{\sqrt{3}+\sqrt{2}}=\frac{(\sqrt{3}-\sqrt{2})}{(\sqrt{3}+\sqrt{2})(\sqrt{3}-\sqrt{2})}=\sqrt{3}-\sqrt{2}$$

$$\frac{1}{\sqrt{5}+\sqrt{4}}=\frac{(\sqrt{5}-\sqrt{4})}{(\sqrt{5}+\sqrt{4})(\sqrt{5}-\sqrt{4})}=\sqrt{5}-\sqrt{4}$$

$$\frac{1}{\sqrt{6}+\sqrt{5}}=\frac{(\sqrt{6}-\sqrt{5})}{(\sqrt{6}+\sqrt{5})(\sqrt{6}-\sqrt{5})}=\sqrt{6}-\sqrt{5}$$

(a) Assume that n is a positive integer. By following the pattern, complete these:

$$\frac{1}{\sqrt{n+1}+\sqrt{n}}=\_\_\_\_\_\_\_\_ \qquad \frac{1}{\sqrt{n+1}-\sqrt{n}}=\_\_\_\_\_\_\_\_$$

(b) Show algebraically that the answer you gave to part (a) is correct.

(c) Find the sum: $\frac{1}{\sqrt{2}+\sqrt{1}}+\frac{1}{\sqrt{3}+\sqrt{2}}+\frac{1}{\sqrt{4}+\sqrt{3}}+\cdots+\frac{1}{\sqrt{2022}+\sqrt{2021}}$.

22 Given the equation $x^3-5x+3=0$:

(a) show that the equation has a root between $x=1$ and $x=2$

(b) show that the equation can be arranged to give $x=\sqrt[3]{5x-3}$

(c) use the iteration $x_{n+1}=\sqrt[3]{5x_n-3}$, with $x_0=1$ to find a solution to the equation, correct to 2 decimal places.

Chapter 2 Function and graphs (I)

2.1 Review of functions

Learning objective

Identify and interpret gradients and intercepts of linear functions, including those involving direct and inverse proportion, graphically and algebraically.

A. Multiple choice questions

1 If (3, 4) is a point on the graph of the inverse proportional function $y = \frac{m}{x}$, then the graph of the function must pass through the point ( ).

A. $(-2, -6)$ B. $(2, -6)$

C. $(4, -3)$ D. $(-4, 3)$

2 If the graph of the function $y = \frac{x}{k}(x < 0)$ passes through the point $(-4, 1)$, then the graph of the function is ( ).

A. a curve B. a straight line

C. a part of curve D. a part of a straight line

3 As shown in the diagram, the graph of a linear function $y = f(x)$, passes through the point $(2, 0)$. If $y > 0$, then the set of values that x can take is ( ).

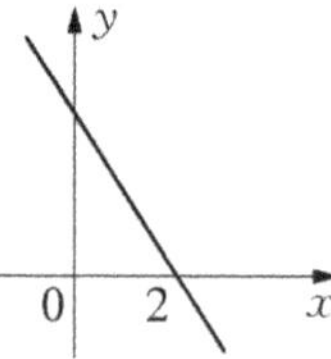

A. $x < 2$ B. $x > 2$

C. $x < 0$ D. $x > 0$

B. Fill in the blanks

4 $A(x_1, y_1)$ and $B(x_2, y_2)$ are two points on the graph of an inverse proportional function $y = \frac{m-3}{x}$. When $x_1 < x_2 < 0$, $y_1 < y_2$, the set of values that m can take is ________.

5 Given that $y = (a - 2)x + a^2 - 9$ represents a proportional function, and the value of y decreases as x increases, then the value of a is ________.

6 The graph of a linear function $y = (m - 1)x + m$ does not pass through the third quadrant. The set of values that m can take is ________.

7 When the line $y = -3x - 2$ is translated 3 units in the positive y-direction, the coordinates of the point of intersection between the resulting line and the x-axis are ________.

C. Questions that require solutions

8 The x-coordinate of the point of intersection of a line l and the line $y = -2x + 7$ is 5. Line l is parallel to the line $y = x - 4$. Find an equation for line l.

9 Given that both the graphs of a direct proportional function $y = kx$ and an inverse proportional function $y = \frac{12}{x}$ pass through $A(4, m)$, answer the following questions.

(a) Find the value of m and an equation for the direct proportion function.

(b) If point P is on the x-axis and the area of $\triangle POA$ is 15 square units, find the coordinates of point P.

10 There are two scales that are commonly used to measure temperature: the Fahrenheit scale (unit: ℉) and the Celsius scale (unit: ℃). The relationship between a temperature in Fahrenheit, y, and the corresponding temperature in Celsius, x, is a linear function. The following table shows some temperatures in degrees Fahrenheit and degrees Celsius.

Temperature, x(℃)	…	0	…	35	…	100	…
Temperature, y(℉)	…	32	…	95	…	212	…

(a) Using the data in the table, write y as a function of x.

(b) A type of thermometer has two scales, where a temperature is shown in degrees Celsius on the left-hand side and in degrees Fahrenheit on the right-hand side. If the Fahrenheit scale reads 56 degrees greater than the Celsius scale, what is the temperature in degrees C?

2.2 Inputs and outputs, inverse functions and composite functions

Learning objective

Interpret simple expressions as functions with inputs and outputs.

A. Multiple choice questions

1 The diagram shows a flowchart that represents a calculation that starts with input x and gives output y. Which of the graphs below shows the relationship of y as a function of x? ( )

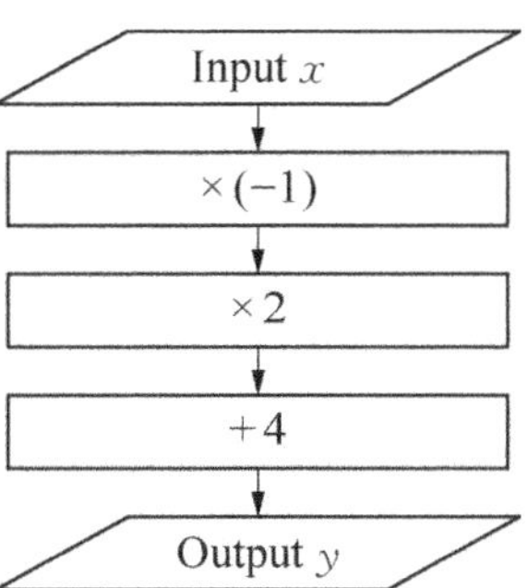

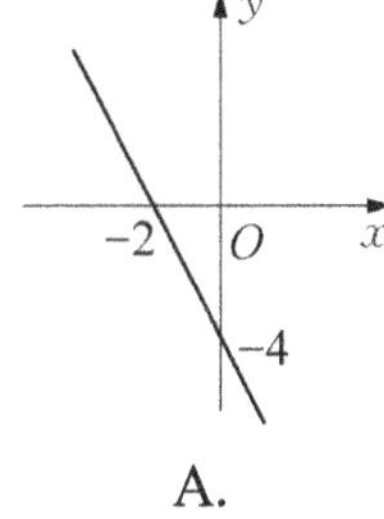

A.

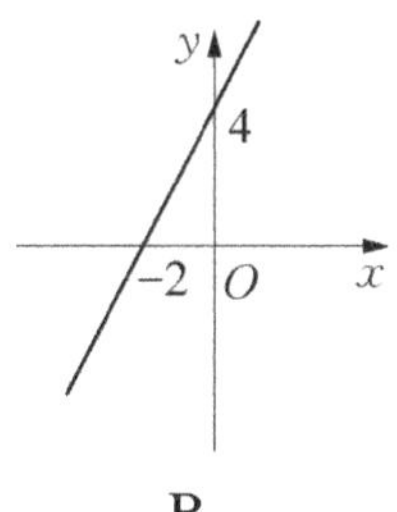

B.

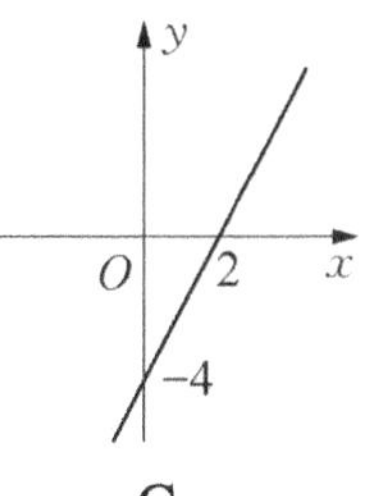

C.

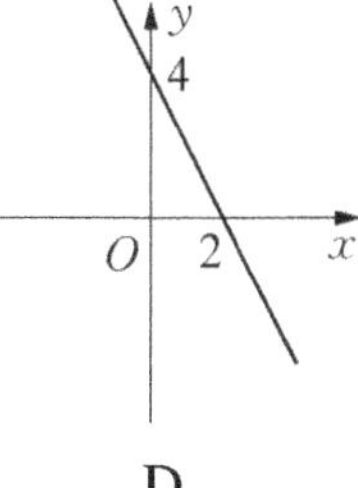

D.

2 If $y + 1$ and $2x$ are in direct proportion, then the function describing the relationship between y and x is ( ).

A. a proportional function
B. a reciprocal function
C. a linear function
D. a quadratic function

3 If y and $x + 1$ are inversely proportional, then the function describing the relationship between y and x is ( ).

A. a direct proportional function
B. an inverse proportional function
C. a linear function
D. none of the above

4 If $f(x) = 2x + 1$, the inverse function of $f(x)$ is ( ).

A. $f(x) = 2x - 1$

B. $f(x) = \frac{2}{x} + 1$

C. $f(x) = \frac{x - 1}{2}$

D. none of the above

B. Fill in the blanks

5 If an equation for a linear function is $y = 2x - 3$, then an equation for x as a function of y is ________.

6 Given that y is a directly proportional function of $\frac{1}{x}$, and when $x = -2$, $y = 3$, then an equation for y as a function of x is ________.

7 If y is a directly proportional function of $x - 1$, then the coordinates of the point that the graph of the function must pass through are ________.

8 Given that y is inversely proportional to $x^2 - 1$, and when $x = 0$, $y = 9$, then, when $y = -3$, $x =$ ________.

9 Based on the calculation shown in the diagram, write an equation for output y as a function of input x: ________.
If $A(x_1, y_1)$ and $B(x_2, y_2)$ are two different points on the graph of the function, and $y_1 = y_2$, then an equation describing the relationship between x_1 and x_2 is ________.

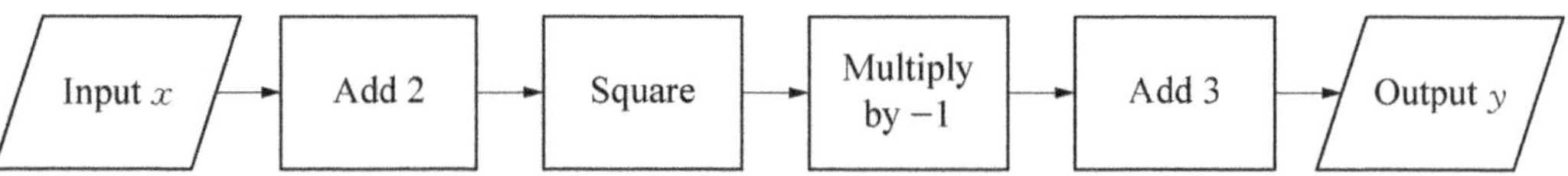

10 Given that $f(x) = 3x - 2$ and $g(x) = x^2$, then:

$fg(2) =$ ________ $ff(2) =$ ________

$gf(2) =$ ________ $gg(2) =$ ________.

C. Questions that require solutions

11 The y-intercept of the graph of a linear function is -3, and when $x = -2$, $y = 3$. Write an equation for the linear function $y = f(x)$. Are y and x in direct proportion?

12 The graph of an inversely proportional function passes through the point $(-2, 4)$.

(a) Find an equation for the inversely proportional function $y = f(x)$.

(b) The graph of $y = f(x)$ is translated 1 unit up, parallel to the y-axis.

(i) Write an equation for the graph of the function obtained.

(ii) Find the inverse of this function.

13 The diagram shows the relationship between the numbers in Circle A and the numbers in Circle B, and the relationship between the numbers in Circle B to the numbers in Circle C.

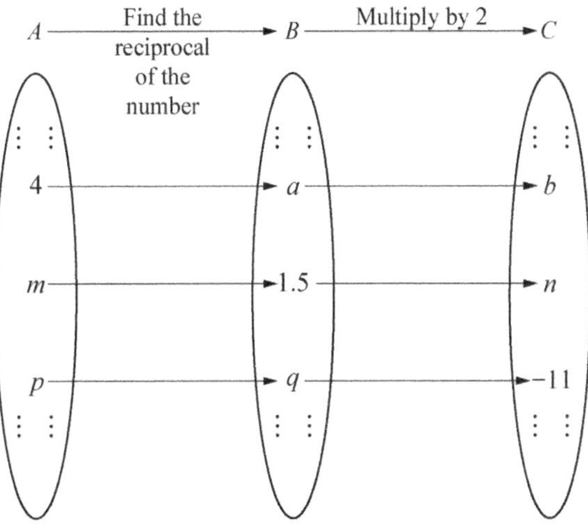

(a) Write the values of:

(i) a and b (ii) m and n (iii) p and q.

(b) Let a number in Circle B be x and the corresponding number in Circle C be y. Using the relationship shown in the diagram, write an equation for y as a function of x.

(c) Let a number in Circle A be x and the corresponding number in Circle C be y. Using the relationship shown in the diagram, write an equation for y as a function of x.

2.3 Further work about linear functions (1)

Learning objective

Interpret simple equations as functions with inputs and outputs and the reverse process as the inverse function.

A. Multiple choice questions

1 Given the two lines $l_1: 3x - 2y + 4 = 0$ and $l_2: x + 2y - 4 = 0$, which of the following statements is correct? ( )

A. The lines are parallel to each other.

B. The lines intersect.

C. The lines are coincident.

D. Their relative position is uncertain.

2 If the line $ax + 2y + 2 = 0$ is parallel to the line $3x - y - 2 = 0$, then the coefficient a is equal to ( ).

A. -3 B. -6 C. $-\frac{3}{2}$ D. $\frac{2}{3}$

3 As shown in the diagram, the gradients of the lines l_1, l_2 and l_3, are k_1, k_2 and k_3, respectively. Therefore ( ).

A. $k_1 < k_2 < k_3$

B. $k_3 < k_1 < k_2$

C. $k_3 < k_2 < k_1$

D. $k_1 < k_3 < k_2$

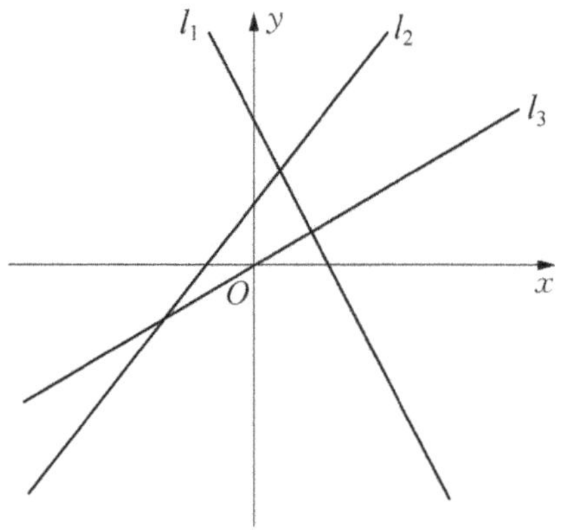

4 If point $A(2, -3)$ is the common point for both lines $a_1x + b_1y + 1 = 0$ and $a_2x + b_2y + 1 = 0$, then an equation for the line that passes through two different points (a_1, b_1) and (a_2, b_2) is ( ).

A. $2x - 3y + 1 = 0$ B. $3x - 2y + 1 = 0$

C. $2x - 3y - 1 = 0$ D. $3x - 2y - 1 = 0$

B. Fill in the blanks

5 A line passing through the points $A(-1, 5)$ and $B(2, 3)$ has a gradient of ________.

6 The equation for the line passing through the point $(2, -1)$ and with a gradient of 3 is ________.

7 The coordinates of the intersection point of the lines $l_1: y = -2x$ and $l_2: y = x + 6$ are ________.

8 The solution of the equation $3x - 1 = -x + 1$ is the x-coordinate of the point of intersection of the line $y = -x$ and the line ________.

9 The line $l: mx + y - 3m + 2 = 0$ must pass through the point with coordinates ________ no matter what value m takes.

10 A line l passes through the point $(-1, 4)$. The intercepts on both the x-axis and y-axis are equal. An equation for the line l is ________.

C. Questions that require solutions

11 On a coordinate grid, draw graphs of the linear functions $y = 2x + 1$ and $y = -x - 2$.

(a) Write the coordinates of the point of intersection.

(b) Use algebra to check your answer to part (a).

12 Use graphs to solve the simultaneous equations $\begin{cases} x + 2y - 3 = 0 \\ 2x - y + 4 = 0 \end{cases}$.

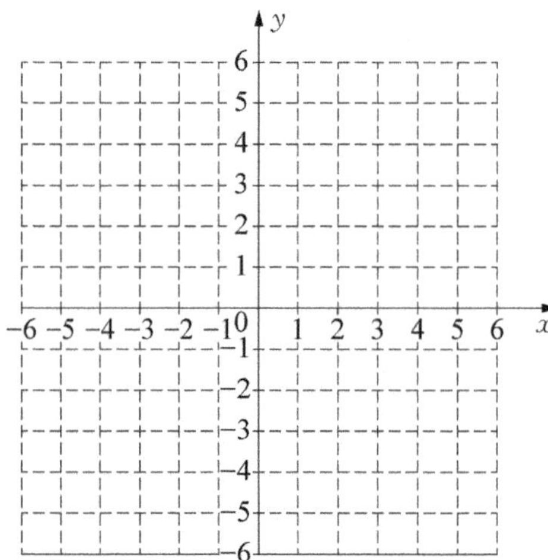

13 Prove that the three points $A(1, -1)$, $B(-2, -7)$ and $C(0, -3)$ are on the same straight line.

2.4 Further work about linear functions (2)

Learning objective

Interpret the succession of two functions as a composite function.

A. Multiple choice questions

1 The correct statement about the relationship between the positions of the lines l_1: $y=-\frac{1}{2}x+3$ and l_2: $x+2y-3=0$ is ( ).

A. They are parallel to each other. B. They intersect.

C. They are coincident. D. It is uncertain.

2 If a line is drawn through the following pairs of points, the line that intersects $y=x-1$ passes through ( ).

A. $(1, 2)$ and $(3, 4)$ B. $\left(-\frac{1}{2}, 1\right)$ and $\left(\frac{3}{2}, 3\right)$

C. $(3, 4)$ and $(6, 8)$ D. $(-2\sqrt{2}, -\sqrt{2})$ and $(2\sqrt{2}, 3\sqrt{2})$

3 Given that a line l passes through the point $(-3, -2)$, that its intercepts on the x-axis and the y-axis are a and b, respectively, and that $a+b=0$, then the line l does not pass through ( ).

A. the first quadrant B. the second quadrant

C. the third quadrant D. the fourth quadrant

4 When using graphs to solve the equation $-\frac{1}{2}x+4=x+1$, you can use the graphs of all the following pairs of lines except ( ).

A. lines $y=-\frac{1}{2}x$ and $y=x-3$ B. lines $y=-x+8$ and $y=2x+2$

C. lines $y=3$ and $y=\frac{3}{2}x$ D. lines $y=\frac{1}{2}x-3$ and $y=x$

B. Fill in the blanks

5 An equation for the line passing through the points $P(2, 1)$ and $Q(-2, 3)$ is ________.

6 Given that a line l with a gradient of $-\frac{1}{3}$ passes through the point $(3, 2)$, the y-intercept of the line l is ________.

7 The equation for the line that passes through the point $A(2, 2)$ and is parallel to the line l: $3x + 4y - 1 = 0$ is ________.

8 If the x-intercept of the line l is -2 and its y-intercept is 3, then the equation for the line l is ________.

9 No matter what value a takes, the coordinates of the point that the line $(a + 1)x + y + 2 - a = 0$ must pass through are ________.

10 Given that a line l with gradient k passes through the point $(-1, 0)$ and intersects the line $y = 2$ in the first quadrant, the set of values that the gradient k can take is ________.

C. Questions that require solutions

11 Two points A and B are on the x-axis, the x-coordinate of point P is 2, and $PA = PB$. If the equation for the line PA is $y = x + 1$, then find an equation for the line PB.

12 Draw graphs to solve the simultaneous equations $\begin{cases} 3x + 4y - 9 = 0 \\ x - 2y - 8 = 0 \end{cases}$.

13 The gradient of a line l is k, the y-intercept of l is -1, and an equation for another line l' is $y = -x + 1$.

(a) If the line l is parallel to the line l', write an equation for the line l.

(b) If the line l intersects the line l', then find the set of values that k can take when the point of intersection is possibly in each quadrant.

2.5 Graphs of functions (1)

Learning objective

Recognise, sketch and interpret graphs of linear functions, quadratic functions and simple cubic functions.

A. Multiple choice questions

1 Given that the graph of the function $y = a(x + m)^2$ is as shown in the diagram, which of the following graphs is closest to the graph of the function $y = mx + a$? (　　)

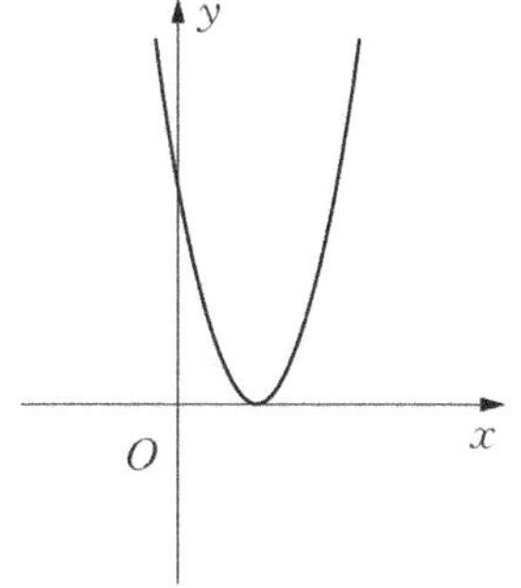

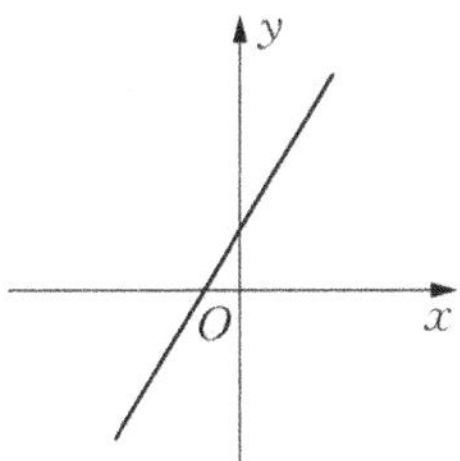

A.

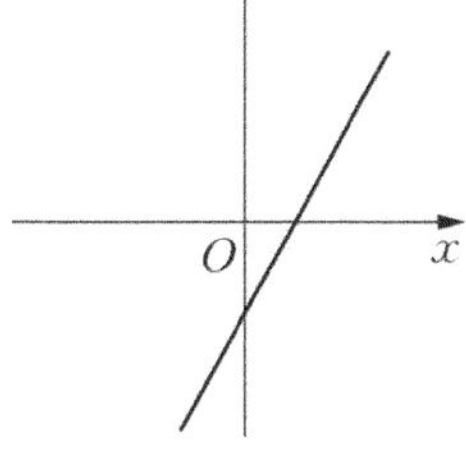

B.

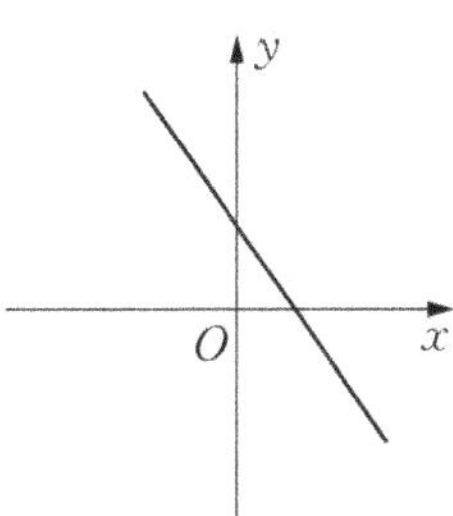

C.

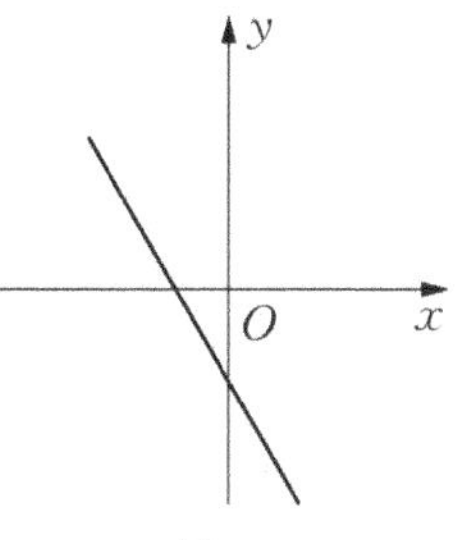

D.

2 The diagram shows the sketches of the graphs of the functions $f(x) = a_1x^3$ and $g(x) = a_2x^3$. Which of the following inequalities is correct? (　　)

A. $0 < a_1 < a_2$　　B. $a_1 < a_2 < 0$

C. $0 < a_2 < a_1$　　D. $a_2 < a_1 < 0$

f(x)　g(x)　y　O　x

3 There are (　　) real solutions to the equation $x^3 - 3x - 1 = 0$.

A. 1　　B. 2

C. 3　　D. 4

4 The diagram shows a cross section of a reservoir, which is divided into a deep-water section and a shallow-water section. If water flows into the reservoir at a constant rate, which of the following graphs best represents the relationship between the maximum depth of the water h at time t? (   )

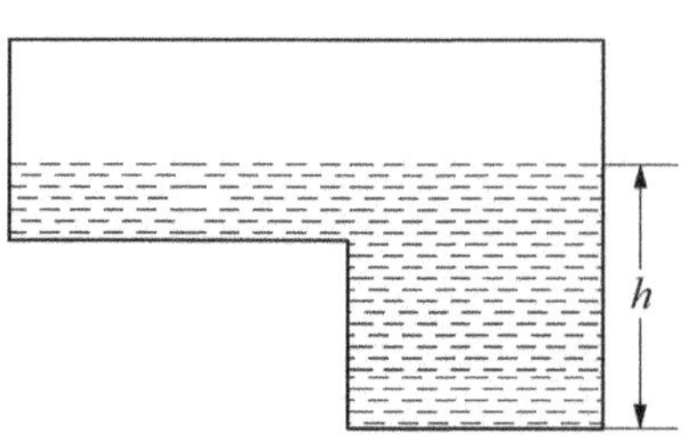

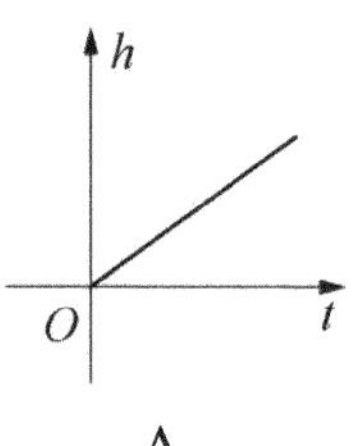

A.

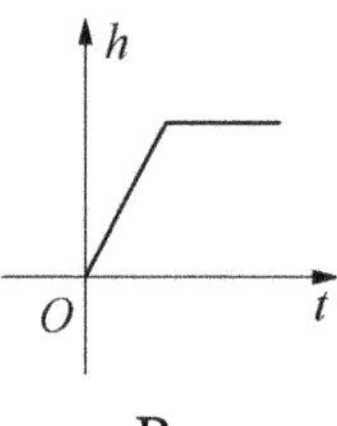

B.

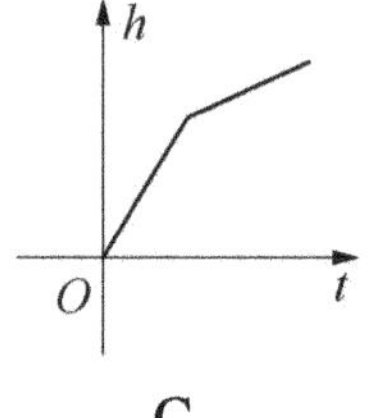

C.

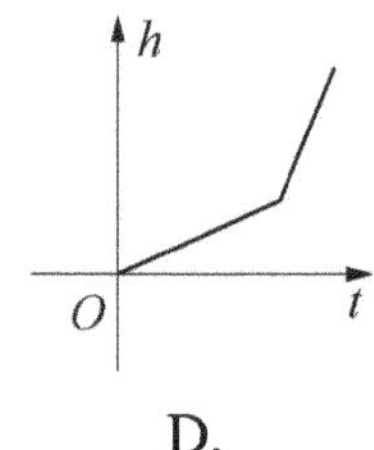

D.

B. Fill in the blanks

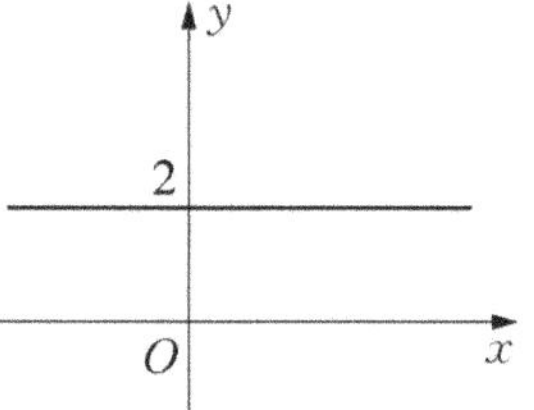

5 The diagram shows the graph of the function $f(x)$. $f(-2) =$ ________.

6 After the line $x + 2y - 2 = 0$ is translated 2 units in the positive x direction, an equation for the resulting line is ________.

7 Given that the x-coordinate of the point of intersection of the lines $l_1: y = ax + 3$ and $l_2: y = 2x + b$ is 1, then $b - a =$ ________.

8 In a Cartesian coordinate system, after point A is translated 2 units in the negative x-direction and then 3 units in the negative y-direction, the resulting point is B. If the equation for the line AB is $y = kx + b$, then $k =$ ________.

9 Given that the equation $-(x - m)^2 + k = 2x + b$ has two real solutions: $-\frac{1}{2}$ and 2, then the solution set to the inequality $-(x - m)^2 + k < 2x + b$ is ________.

10 The solution set to the inequality $x^3 - x^2 < 0$ is ________.

C. Questions that require solutions

11 Draw the graphs of the function $y = -(x-1)^2 + 2$ and the line $x + y - 1 = 0$ on the same Cartesian coordinate system.

12 By drawing suitable graphs, find the solution to the equation $x^3 + 3x^2 - 4 = 0$.

13 A patrol boat and a cargo vessel depart at the same time from Port A and sail to Port B. The distance between ports A and B is 100 kilometres. The speeds of the patrol boat and the cargo vessel are 100 km/h and 20 km/h, respectively. The patrol boat travels continually back and forth between the two ports (ignore the time taken for each turn). How many times does the cargo vessel pass the patrol boat between its departure from Port A and its arrival at Port B? (You may solve the problem by drawing graphs of suitable functions.)

2.6 Graphs of functions (2)

Learning objective

Find the equation of the line through two given points, or through one point with a given gradient and solve problems involving direct and inverse proportion.

A. Multiple choice questions

1 Given that the y-value of the function $y = kx$ decreases as x increases, which of the following is a rough sketch of the graph of this function and the graph of the function $y = \frac{k}{x}$ on the same Cartesian coordinate system? (　　)

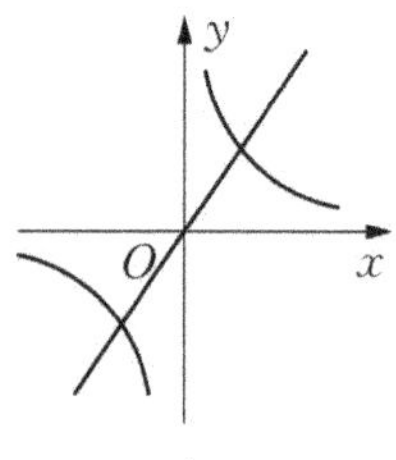

A.

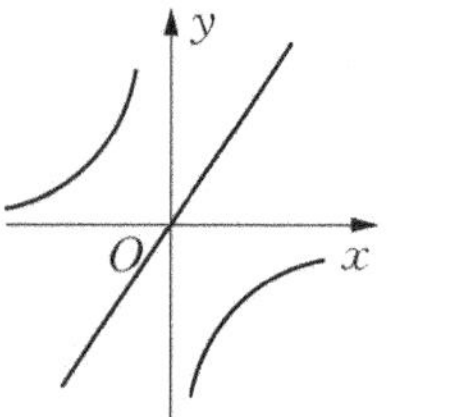

B.

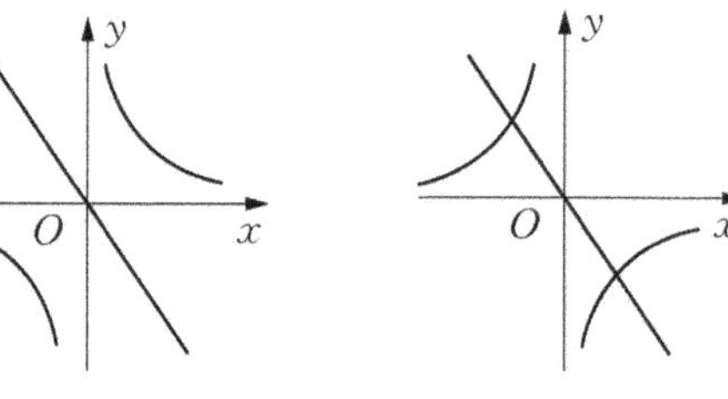

C.　　D.

2 Given that x is directly proportional to $\frac{3}{y}$ and y is inversely proportion to $2z$, then an equation for x as a function of z is (　　).

A. $xz = 6$

B. $x = \frac{3}{2}z$

C. $x = kz(k \neq 0)$

D. $x = \frac{k}{z}(k \neq 0)$

3 The diagram shows part of the graphs of the three above the x-axis. The relationship between k_1, k_2 and k_3 is (　　).

A. $k_1 > k_2 > k_3$

B. $k_3 > k_2 > k_1$

C. $k_2 > k_3 > k_1$

D. $k_3 > k_1 > k_2$

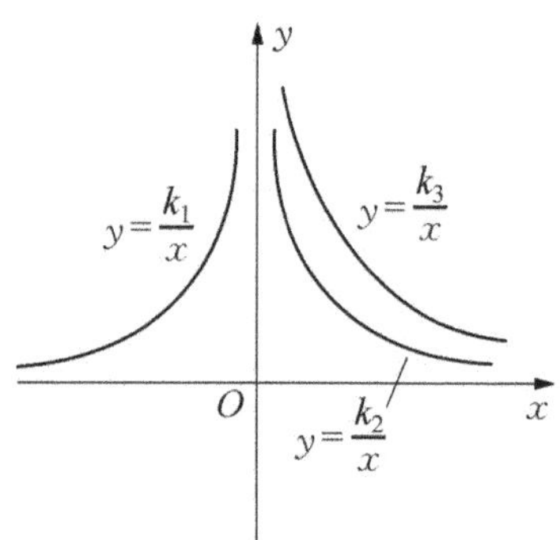

4 The set of values that the solution to the equation $x^3 + x - 1 = 0$ can most likely take is ( ).

A. $0 < x < 1$

B. $\frac{1}{2} < x < 1$

C. $\frac{3}{5} < x < 1$

D. $\frac{3}{5} < x < \frac{4}{5}$

B. Fill in the blanks

5 Given that the coordinates of one of the points of intersection of the graphs of a directly proportional function and an inversely proportional function are $(2, -4)$, then the coordinates of the other point of intersection of the graphs are ________.

6 The diagram shows the graph of an inversely proportional function in the second quadrant. Point A is a point on the graph, AM is perpendicular to the x-axis and M is the foot of the perpendicular. If the area of triangle AOM is 3 square units, then an equation for the inversely proportional function is ________.

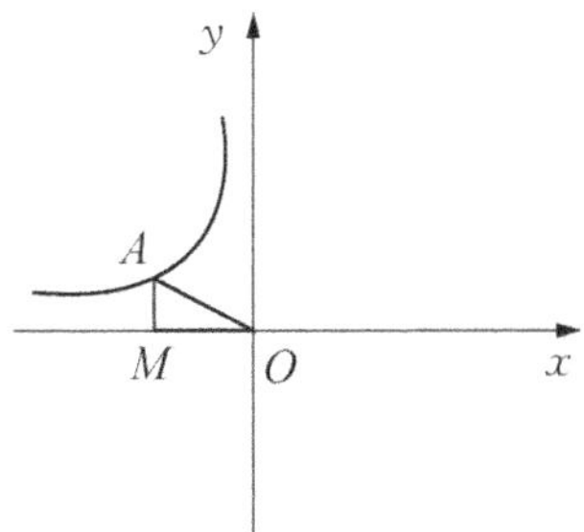

7 The graphs of the functions $y = x^3$ and $y = 2x^2 - 1$ have ________ point(s) of intersection.

8 When the line $y = k_1x + b$ and the graph of the inversely proportional function $y = \frac{k_2}{x}$ intersect at $A(1, 6)$ and $B(a, 3)$, then the solution set to the inequality $k_1x + b - \frac{k_2}{x} > 0$ is ________.

9 The equation $2^x + x^2 = \sqrt{2}$ has ________ real solution(s).

10 If a, b and c are positive, the real roots of the equations $2^x = 5$, $x^2 = 5$ and $\frac{5}{11}x + 4 = 5$, respectively, then comparing the values of a, b and c, ________ < ________ < ________. (Hint: you may use the graphs to find the answer.)

C. Questions that require solutions

11 The graphs of the linear function $y_1 = -x + b$ and the inversely proportional function $y_2 = \frac{k}{x}$ intersect at points $A(5, 1)$ and A'.

(a) Find algebraic expressions for the two functions.

(b) Using the graphs of the functions, write the set of values that the variable x can take when $y_1 < y_2$.

12 By drawing graphs, find the number of points of intersection of the line $y = x - 1$, the hyperbola $y = \frac{2}{x}$ and the parabola $y = -2x^2 + 12x - 15$.

13 By drawing graphs, determine the number of real solutions to the equation $x^3 + 2x^2 - 2x - 2 = 0$.

Unit test 2

A. Multiple choice questions

1 As shown in the diagram, the graph of a linear function, $y = -2x + m$, passes through the point $(2, 0)$. The value that m can take is (　　).

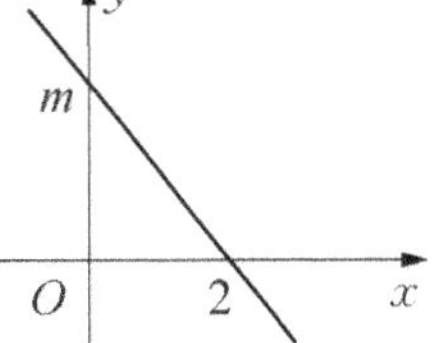

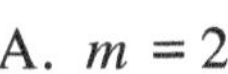

A. $m = 2$　　B. $m = -2$

C. $m = -4$　　D. $m = 4$

2 When the line $y = 2x + 1$ is translated 2 units down parallel to the y-axis, the coordinates of the point of intersection of the resulting line and the y-axis are (　　).

A. $(-1, 0)$　　B. $(1, 0)$

C. $(0, 1)$　　D. $(0, -1)$

3 Given the two lines l_1: $y = 2x - 1$ and l_2: $y = -2x + 3$, which of the following statements is correct? (　　)

A. The lines are parallel.　　B. The lines intersect.

C. The lines are coincident.　　D. The relative position is uncertain.

4 Given that a line l passes through the point $(-1, 2)$, its intercept on the x-axis is point $(a, 0)$, and $a > 0$, then the line l must not pass through (　　).

A. the first quadrant　　B. the second quadrant

C. the third quadrant　　D. the fourth quadrant

5 Given that the graph of a function $y = a(x + m)^2$ is as shown in the diagram, which of the following statements is correct about the line $y = mx + a$? (　　)

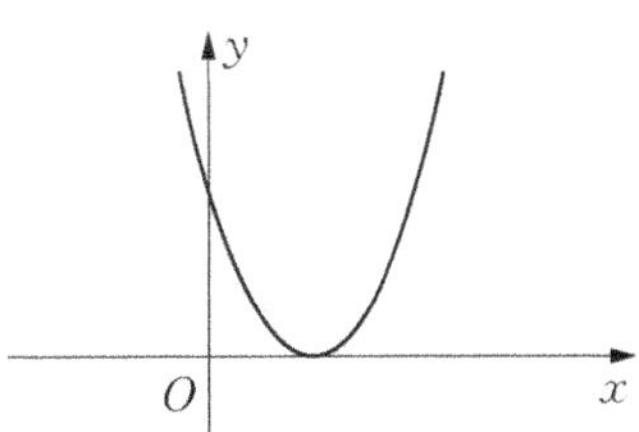

A. When x increases, y increases.

B. When x increases, y decreases.

C. The line goes through the third quadrant.

D. None of the above are correct.

6 Given that, for the function $y = kx$, the value of y increases as x increases, then a sketch of the graph of $y = kx$ and the graph of the function $y = \frac{k}{x}$ on the same Cartesian coordinate system could be ( ).

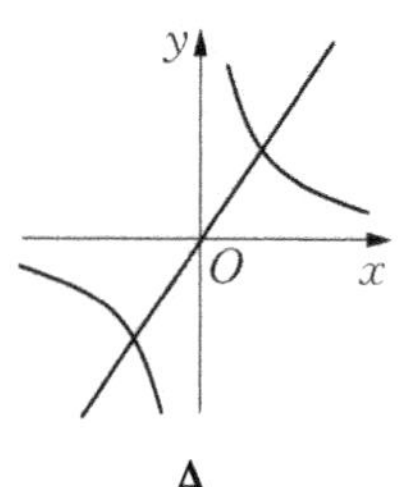

A.

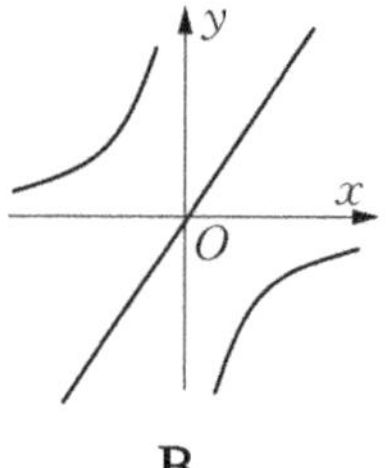

B.

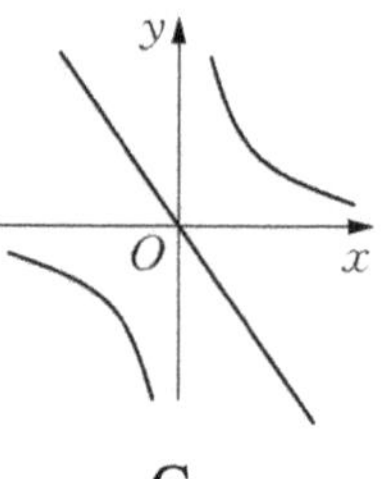

C.

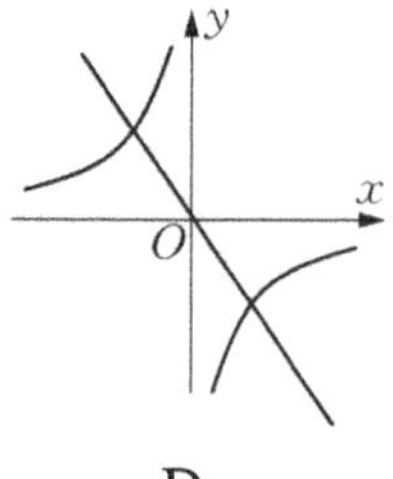

D.

B. Fill in the blanks

7 An equation for the line passing through the points (1, −1) and (−1, 1) is ________.

8 $A(1, 2)$ is a point on the graph of an inversely proportional function $y = \frac{k^2 - 1}{x}$, where k is a constant. The values that k can take are________.

9 y is inversely proportional to $3x + 1$. When $x = 0$, $y = 2$. So when $x = -1$, $y =$______.

10 An equation for the line that passes through the point $A(-1, 3)$ and is parallel to the line l: $3x + 2y - 1 = 0$ is ________.

11 Two points $A(-3, 6)$ and $B(3, -2)$ are both on the graph of the linear function l: $y = (t - 1)x + m$. Therefore an equation for the line l is ________.

12 Given that $f(x) = 2x + 1$ and that $g(x) = (x + 1)^2$, then:

$gf(1) =$ ________

$fg(1) =$ ________

$ff(1) =$ ________

$gg(1) =$ ________.

13 After the line $3x + y - 4 = 0$ is translated 3 units in the positive x-direction, the equation for the resulting line is ________.

14 The line l: $(2m - 1)x - (m + 3)y - (m - 1) = 0$ must pass through the point with coordinates ________ no matter what value m takes.

15 The line l: $y = -1$ has no points of intersection points with the curve $(x-2)^2 + m - y = 0$. The set of values that m can take is ________.

16 The graphs of the functions $y = x^3 - 2$ and $y = 2x + 1$ have ________ point(s) of intersection.

C. Questions that require solutions

17 The diagram shows the lines l and m.

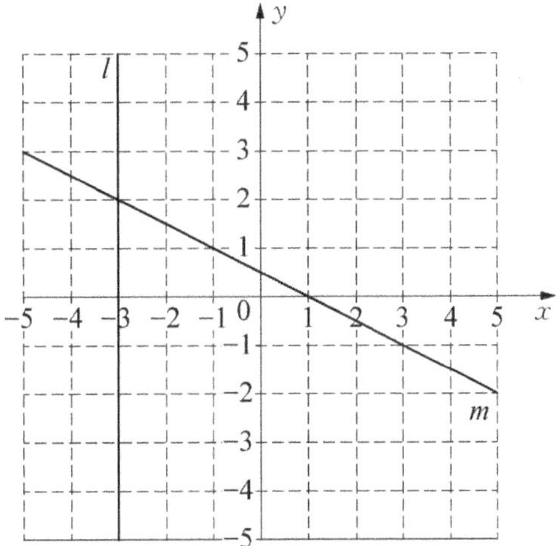

(a) Write an equation for the line m.

(b) Write the coordinates of the point of intersection of lines l and m.

(c) Find the gradient of the line m.

(d) The graph of the function $y = -(x - 1)^2 + 2$ and the line m have ________ points of intersection. (Hint: you may use the graphs to find the answer.)

18 You are given $f(x) = 2x + 3$, $g(x) = \sqrt{x}$ and $h(x) = x^2 + 1$.

(a) If $f(x) = 7$, find the values of $g(x)$ and $h(x)$.

(b) Find $fg(2)$.

(c) Find the inverse function of $f(x)$.

(d) Find the value of x which satisfies $hf(x) = 5$.

19 Two lines $l_1: 2x + y - 2 = 0$ and $l_2: x - 2y + 4 = 0$ intersect at point Q.

(a) Find the coordinates of point Q.

(b) The gradient of line l_3 is 2, and the graph of l_3 passes through point Q. Find an equation for line l_3.

Chapter 3 Function and graphs (II)

3.1 Translations and reflections of the graph of a given function

Learning objective

Sketch and interpret translations and reflections of the graph of a given function.

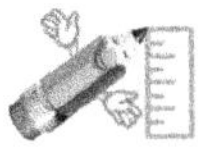

A. Multiple choice questions

1 Which of the following statements about the inverse proportional function $y = \frac{3}{x}$ is incorrect? (   )

A. The graph of the inverse proportional function $y = \frac{3}{x}$ is symmetrical about the origin.

B. After reflecting the graph of the inverse proportional function $y = \frac{3}{x}$ in the x-axis, an equation of the graph obtained is $y = -\frac{3}{x}$.

C. After reflecting the graph of the inverse proportional function $y = \frac{3}{x}$ in the y-axis, an equation of the graph obtained is $y = -\frac{3}{x}$.

D. The value y of the inverse proportional function $y = -\frac{3}{x}$ decreases as x increases.

2 After the parabola C: $y = x^2 + 2x - 3$ is translated, the resulting parabola is C'. If the two parabolas are symmetrical about the line $x = 1$, then which of the following statements is correct? (   )

A. By translating the parabola C to the right by $\frac{5}{2}$ units along the x-axis, the resulting parabola is C'.

B. By translating the parabola C to the right by 4 units along the x-axis, the resulting parabola is C'.

C. By translating the parabola C to the right by $\frac{7}{2}$ units along the x-axis, the resulting parabola is C'.

D. By translating the parabola C to the right by 6 units along the x-axis, the resulting parabola is C'.

3 Which of the following is/are the graph(s) of a linear function(s)? (   )

I

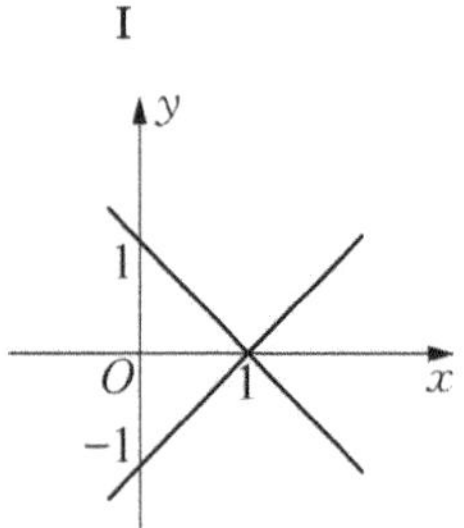

II

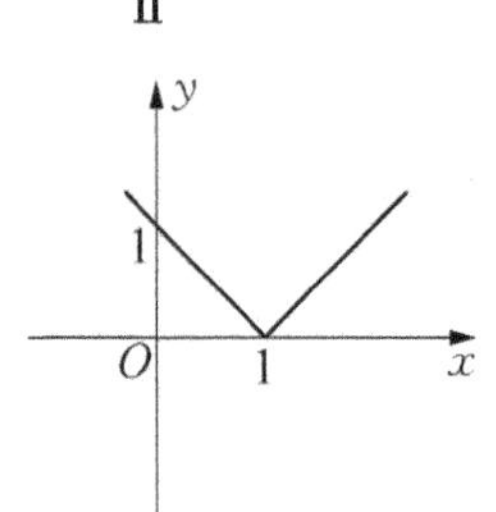

III

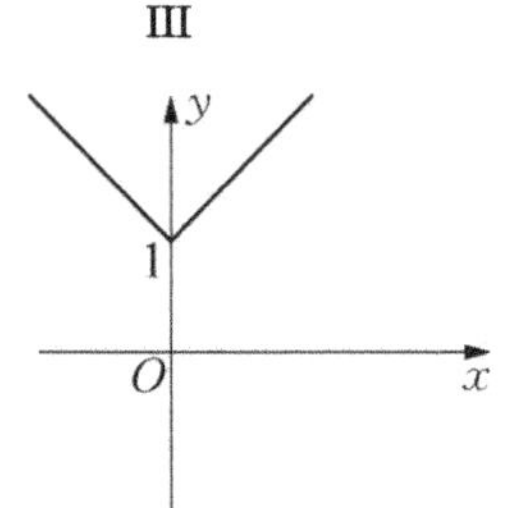

IV

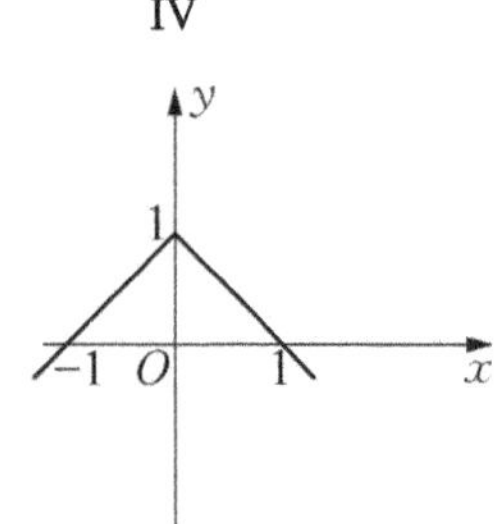

A. I　　B. I and II

C. II, III and IV　　D. none of them

4 If the graph of $y = x^2$ is translated $2\sqrt{2}$ units along the line $y = x$, and the turning point of the resulting parabola is at point A on the line $y = x$, then an equation of the resulting parabola is: (   ).

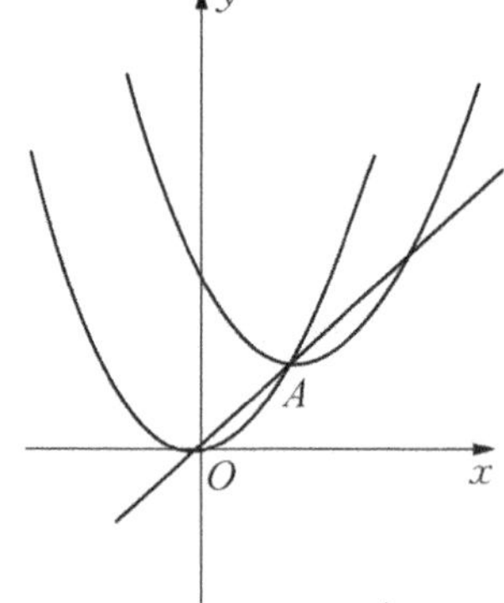

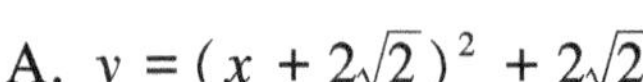

A. $y = (x + 2\sqrt{2})^2 + 2\sqrt{2}$　　B. $y = (x + 2)^2 + 2$

C. $y = (x - 2\sqrt{2})^2 + 2\sqrt{2}$　　D. $y = (x - 2)^2 + 2$

B. Fill in the blanks

5 After reflecting the line $y = -\frac{1}{2}x - 3$ in the y-axis, an equation of the resulting line is ________.

6 After translating the graph of the linear function $y = 2x - 4$ ________ units to the ________ along the x-axis, an equation of the resulting graph, which is a proportional function, is ________.

7 After reflecting the parabola $y = -\frac{1}{2}x^2 + x - 3$ in the x-axis, an equation of the resulting parabola is ________.

8 Given that after translating the line $y = kx + b$ to the right by 3 units, and then down by 2 units, the resulting line coincides with the original line, then the value of k is ________.

9 After reflecting the line $y = -3x + 4$ in the line $y = -3x + 1$, an equation of the resulting line is ________.

10 Write two properties of the function $y = \frac{1}{x^2}$.

(a) ______________________________

(b) ______________________________.

C. Questions that require solutions

11 The equation of a straight line is $y = \frac{3}{2}x - 6$.

(a) The line is reflected in the x-axis. What is the equation of the reflected line?

(b) If the line $y = kx$ intersects both the line $y = \frac{3}{2}x - 6$ and its reflection, find the set of values that k can take.

12 Given that the line $l: x - 3y - 9 = 0$ and the line l' are symmetrical about the line $y = x$, find the equation of the line l'.

13 Sketch the graph of the function $y = x^2 - 6x + 5$. Determine the number of solutions to the equation $x^2 - 6x + 5 = k$, when k takes different values.

3.2 Graphs of quadratic functions $y=a(x+m)^2+k$ (1)

Learning objective

Identify and interpret roots, intercepts and turning points of quadratic functions, deducing roots algebraically and graphically.

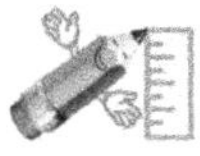

A. Multiple choice questions

1 If the quadratic function $y=\frac{1}{2}x^2+2x+1$ is written in the form of $y=a(x+m)^2+k$, it becomes (　　).

A. $y=\frac{1}{2}(x-1)^2+2$　　B. $y=\frac{1}{2}(x+1)^2+\frac{1}{2}$

C. $y=\frac{1}{2}(x-1)^2-3$　　D. $y=\frac{1}{2}(x+2)^2-1$

2 The coordinates of the turning point of the quadratic function $y=-3x^2-6x+5$ are (　　).

A. $(-1, 8)$　　B. $(1, 8)$

C. $(-1, 2)$　　D. $(1, -4)$

3 To obtain the graph of the quadratic function $y=-x^2+2x-2$, the graph of $y=-x^2$ should be translated (　　).

A. 2 units in the negative x-direction, and then 2 units in the negative y-direction

B. 2 units in the positive x-direction, and then 2 units in the positive y-direction

C. 1 unit in the negative x-direction, and then 1 unit in the positive y-direction

D. 1 unit in the positive x-direction, and then 1 unit in the negative y-direction

4 The graph shows the quadratic function $y=ax^2+bx+c$. Which of the following statements is/are correct? (　　)

① $a>0$.

② The graph is symmetrical with respect to the line $x=1$.

③ When $x=-1$ or $x=3$, $y=0$.

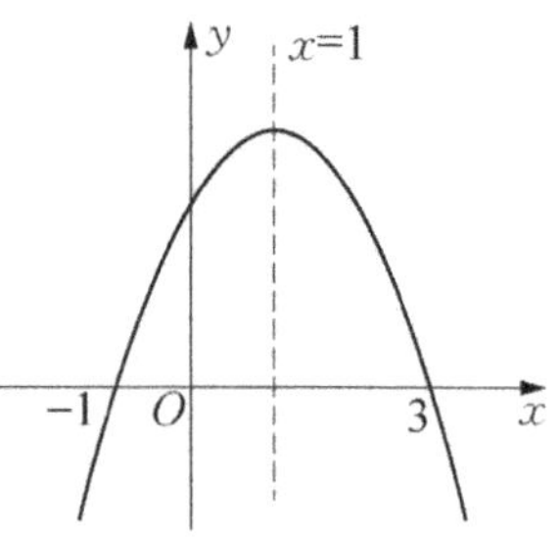

A. ① and ②　　B. ② and ③

C. ① and ③　　D. ①, ② and ③

B. Fill in the blanks

5 The line of symmetry of the parabola $y = 2(x + 1)^2 - 3$ is ________.

6 The coordinates of the turning point of the graph of the quadratic function $y = \frac{1}{2}x^2 - 6x + 15$ are ________.

7 Given a quadratic function $y = \frac{1}{2}x^2 - 6x + 15$, when $x =$ ________, y has its maximum/minimum value, which is ________. (Select which option, maximum or minimum, is correct.)

8 Given that point M is on the parabola $y = x^2 + 2x - 1$, and its coordinate on the x-axis is −3, if point N is symmetrical to point M with respect to the line of symmetry of the parabola, then the coordinates of point N are ________.

9 After translating the parabola $y = x^2 + 2x - 1$ by 2 units in the positive x-direction and 1 unit in the positive y-direction, the equation of the resulting parabola is ________.

10 If the parabola $y = mx^2 + 2x + m - 4m^2$ passes through the origin, then the coordinates of the turning point of the parabola are ________.

C. Questions that require solutions

11 Given the parabola $y = -2x^2 + 4x + 6$, identify whether it is U-shaped or an inverted U-shape, find the equation of the line of symmetry and the coordinates of the turning point, and then sketch the parabola.

12 The graph of a quadratic function $y = -2x^2 + bx + c$ passes through points $A(0, 4)$ and $B(1, -2)$.

(a) Find the equation of the quadratic function.

(b) Write the equation in the form of $y = a(x + m)^2 + k$, and then write the coordinates of the turning point C of the parabola.

13 Given that the line of symmetry of a quadratic function $y = ax^2 + bx + c$ is the line $x = 2$, the y-intercept of its graph is 6, and the equation $ax^2 + bx + c = 0$ has a solution $x = 1$, find the equation of the quadratic function, and write down the coordinates of the turning point of its graph.

3.3 Graphs of quadratic functions $y=a(x+m)^2+k$ (2)

Learning objective

Identify and interpret roots, intercepts and turning points of quadratic functions, deducing roots algebraically (completing the square) and graphically.

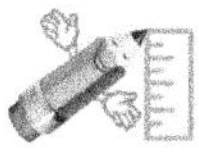

A. Multiple choice questions

1 The graph of a quadratic function $y=-x^2+bx+c$ has a maximum value at $(1, -4)$. The values of b and c are ( ).

A. $b=2$, $c=5$

B. $b=-2$, $c=-5$

C. $b=-2$, $c=5$

D. $b=2$, $c=-5$

2 Read the following statements about the quadratic function $y=ax^2+bx+c$.

① When $c=0$, the graph of the function passes through the origin.

② When $b=0$, the graph of the function is symmetrical with respect to the y-axis.

③ The highest value of the y-coordinate of the graph of the function is $\frac{4ac-b^2}{4a}$.

④ When $c>0$ and the graph of the parabola $y=ax^2+bx+c$ is an inverted U-shaped, the equation $ax^2+bx+c=0$ must have two unequal real roots.

How many of these statements is/are true? ( )

A. one

B. two

C. three

D. four

3 If the line of symmetry of the graph of the quadratic function $y=4x^2-mx+5$ is the line $x=-2$, then when $x=1$, the value of y is ( ).

A. -7

B. 1

C. 17

D. 25

4 The diagram shows the graph of a quadratic function $y = ax^2 + bx + c$. How many of the following conclusions is/are correct? (　　)

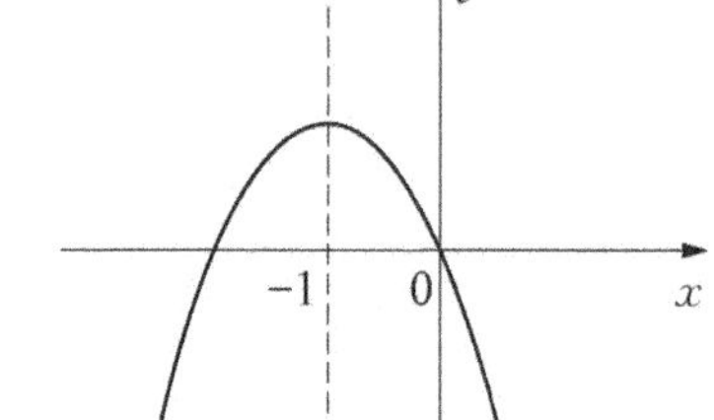

① $a + b + c > 0$ ② $a - b + c > 0$ ③ $abc < 0$ and ④ $2a - b = 0$.

A. one

B. two

C. three

D. four

B. Fill in the blanks

5 If the y-axis is the line of symmetry of the parabola $y = -\frac{1}{2}x^2 + (m + 1)x - m + 2$, then the coordinates of the turning point of this parabola are ________.

6 Given that the turning point of a parabola $y = x^2 - x + m$ is above the x-axis, then the set of values that m can take is ________.

7 Given that the x-coordinate of the turning point of the graph of a quadratic function $y = ax^2 - (a + 1)x - 2$ is 1, then when $x < 1$, the value of y ________ as the value of x increases, and when $x > 1$, the value of y ________ as the value of x increases.

8 Given that the quadratic equation $ax^2 + bx + c + 1 = 0$ has two real roots, $x = -1$ and $x = 5$, then the line of symmetry of the graph of the quadratic function $y = ax^2 + bx + c$ is ________.

9 Given the quadratic function $y = ax^2 - 2x + 2 (a > 0)$, then its graph must not pass through the ________ quadrant.

10 Given that the line of symmetry of the parabola $y = ax^2 + bx + c$ is the line $x = 2$, and the parabola passes through the point $(3, 0)$, then the value of $a + b + c$ is ________.

C. Questions that require solutions

11 Write the parabola $y = -\frac{1}{2}x^2 - 3x - \frac{5}{2}$ in the form of $y = a(x + m)^2 + k$ by completing the square.

(a) Write the coordinates of the turning point and the line of symmetry of the parabola, and sketch the parabola.

(b) According to the graph, write the values that x can take so that $y > 0$.

12 Given that the line of symmetry of the graph of the quadratic function $y = (m^2 - 2)x^2 - 4mx + n$ is the line $x = 2$, and the graph has a maximum turning point at the line $y = \frac{1}{2}x + 1$, find the equation of the quadratic function.

13 Consider the quadratic function $y = x^2 + mx + m - 5$.

(a) Prove that the graph of the function $y = x^2 + mx + m - 5$ always has two points of intersection with the x-axis for whatever value that m takes.

(b) What value can m take when the two points of intersection of the graph of $y = x^2 + mx + m - 5$ and the x-axis are the shortest distance apart?

3.4 Graphs of quadratic functions $y=a(x+m)^2+k$ (3)

Learning objective

Identify and interpret roots, intercepts and turning points of quadratic functions, deducing roots algebraically and graphically. Find approximate roots to simultaneous equations (linear and quadratic) graphically.

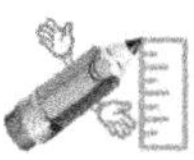

A. Multiple choice questions

1 The condition for the turning point of a parabola $y=ax^2+bx+c$ $(a<0)$ to be above the x-axis is (   ).

A. $b^2-4ac>0$ B. $b^2-4ac<0$ C. $b^2-4ac\geqslant 0$ D. $b^2-4ac\leqslant 0$

2 The graph of a quadratic function $y=ax^2+bx$ is as shown in the diagram. Then the values that a and b can take are (   ).

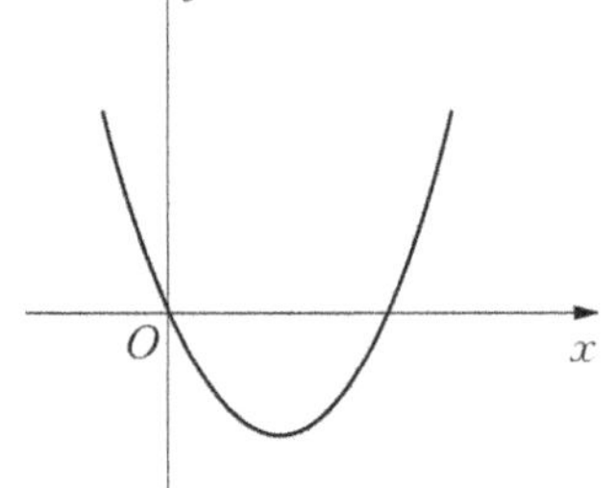

A. $a>0$, $b>0$

B. $a<0$, $b>0$

C. $a>0$, $b<0$

D. $a<0$, $b<0$

3 The diagram shows part of a parabola $y=-ax^2+bx+c$. If $y>0$, then the set of values that x can take is (   ).

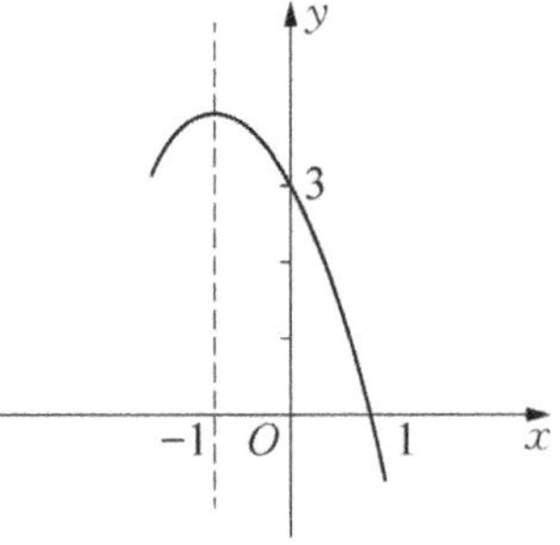

A. $-4<x<1$

B. $-3<x<1$

C. $x<-4$ or $x>1$

D. $x<-3$ or $x>1$

4 Given a linear function $y=ax+b$ and a quadratic function $y=ax^2+bx+c(a\neq 0)$, their graphs in the same Cartesian coordinate system may be (   ).

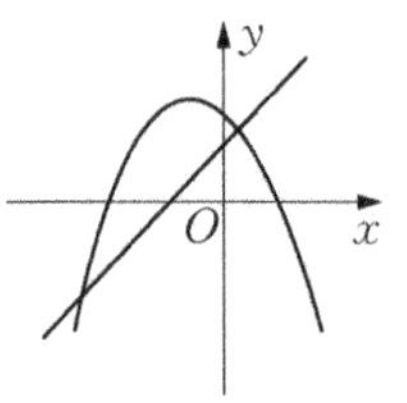

A.

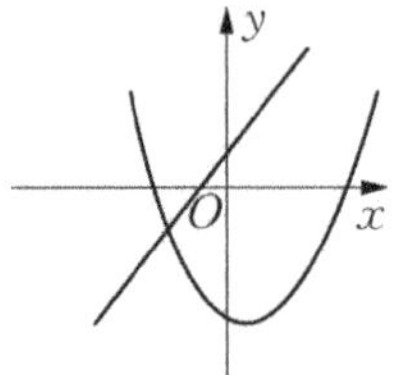

B.

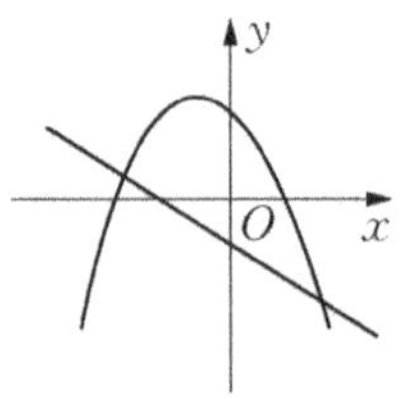

C.

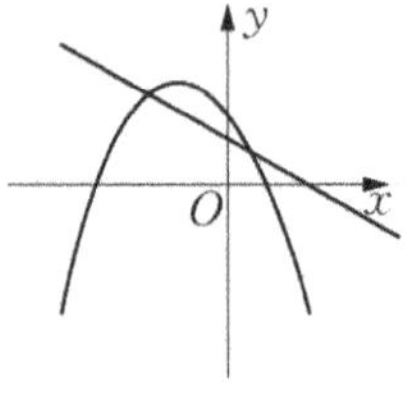

D.

B. Fill in the blanks

5 If the graph of a quadratic function $y = (m-1)x^2 + x + m^2 - 1$ passes through the origin, then the value of m is ________.

6 The coordinates of the turning point of the parabola that passes through the three points $(-1, 0)$, $(3, 0)$ and $(1, 2)$ are ________.

7 A parabola $y = ax^2$ is translated with vector $\begin{pmatrix} 2 \\ 3 \end{pmatrix}$. If the resulting parabola passes through the point $(3, -1)$, then the equation for the resulting parabola is ________.

8 If the parabola $y = x^2 - bx + 4$ has one and only one point of intersection with the x-axis, then the value(s) of b is/are ________.

9 If the graph of a quadratic function $y = ax^2 + bx + c$ $(a > 0)$ passes through points $A(-1, y_1)$ and $B(2, y_2)$, and the two real roots of the equation $ax^2 + bx + c = 0$ are $x = -2$ and $x = 4$, then y_1 ________ y_2. (Fill in with ">", "<" or "=".)

10 The solutions to the equation $x^2 + x - 2 = 0$ can be considered to be the x-coordinates of the points of intersection between the parabola $y = x^2 + x - 2$ and the x-axis, and they can also be considered to be the x-coordinates of the points of intersection points between the parabola $y = x^2$ and the line ________.

C. Questions that require solutions

11 The graph of a quadratic function $y = x^2 + bx + c$ passes through points $A(2, 2)$ and $B(5, 2)$.

(a) Find the values of b and c.

(b) Let the points of intersection points between the graph of this function and the x-axis be points C and D (C is to the right of D). Find the area of the quadrilateral $ABCD$.

12 As shown in the diagram, the graph of a quadratic function $y = ax^2 + 4x + c$ is U-shaped, and it intersects the x-axis at the points A and B (point A is to the left of point B). It intersects the y-axis at point C, where $AB = 2$ and $OA = OC$.

(a) Find the equation of the quadratic function.

(b) After translating the graph 3 units to the right, and m unit(s) down, the resulting graph intersects the y-axis at the point $(0, -1)$. Find the equation of the resulting graph.

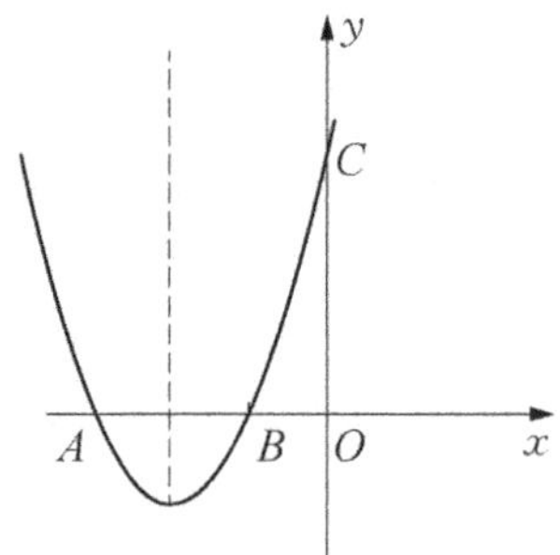

13 Consider the parabola $y = x^2$ and a line $y = (m^2 - 1)x + m^2$.

(a) For what value of m, are there two points of intersection between the parabola and the line?

(b) Let the intersection points of the parabola and the line be A and B, from left to right. When the difference between the x-coordinates of these two points of intersection is 3, find the height on OB in triangle AOB (where O is the origin).

3.5 Graphs of quadratic functions $y=a(x+m)^2+k$ (4)

Learning objective

Identify and interpret roots, intercepts and turning points of quadratic functions, deducing roots algebraically and graphically; find approximate roots to simultaneous equations (linear and quadratic) graphically.

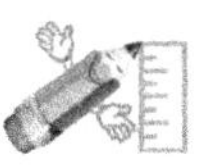

A. Multiple choice questions

1 In the Cartesian coordinate system, if the parabola $y=-(x-1)^2$ is translated 4 units in the negative x-direction, then the equation of the resulting parabola is ( ).

A. $y=-(x-5)^2$

B. $y=-(x+3)^2$

C. $y=-(x-1)^2+4$

D. $y=-(x-1)^2-4$

2 If a line $y=ax+b$ passes through the first, second and fourth quadrants, then the turning point of the parabola $y=ax^2+bx$ is in ( ).

A. the first quadrant

B. the second quadrant

C. the third quadrant

D. the fourth quadrant

3 Given that, for the inverse proportional function $y=\frac{a}{x}$, when $x>0$, the value of the function y decreases as x increases, then the graph of the quadratic function $y=ax^2-ax$ may be ( ).

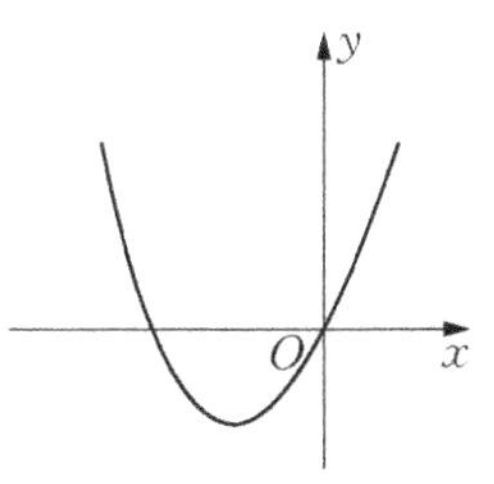

A.

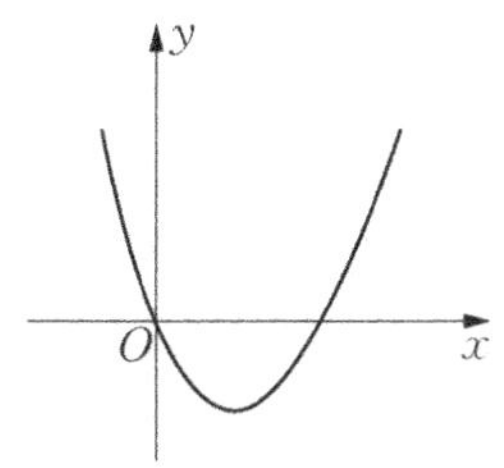

B.

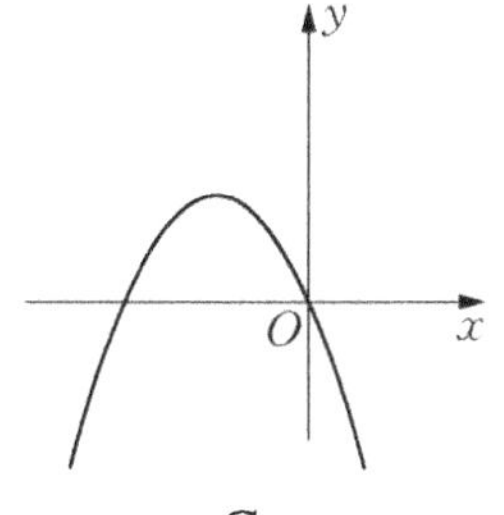

C.

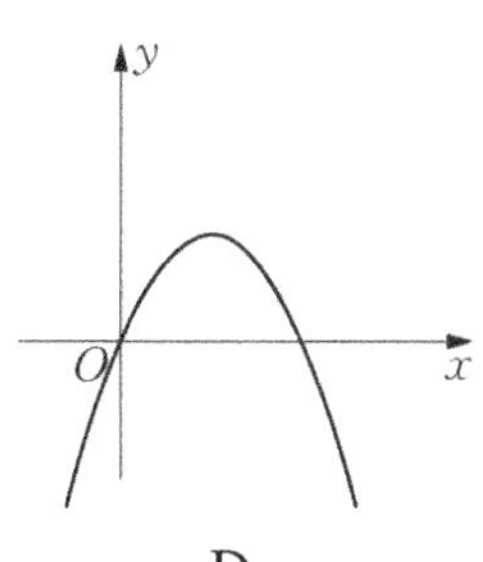

D.

4 Given that the graph of a quadratic function $y = ax^2 + bx + c$ is as shown in the diagram, how many of the following conclusions is/are correct? (　　)

① $abc < 0$　　② $a + c < b$

③ $a + b + c > 0$　　④ $2c < 3b$

A. one　　B. two

C. three　　D. four

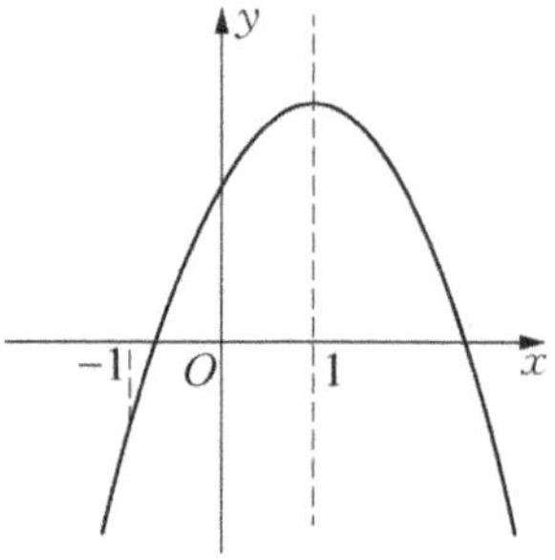

B. Fill in the blanks

5 If the graph of a quadratic function $y = (a - 2)x^2 - 1$ does not pass through the second quadrant, then the set of values that a can take is ________.

6 The turning point of the parabola $y = ax^2 + bx + c$ ($a > 0$, $b > 0$, $c < 0$) is in the ________ quadrant.

7 If the line of symmetry of the graph of a quadratic function $y = ax^2 + bx + c$ is the line $x = 2$, and the minimum value of the function is -2, then the real solutions to the equation $ax^2 + bx + c + 2 = 0$ in x are ________.

8 Given that the graph of a quadratic function is U-shaped, the equation of its line of symmetry is $x = -1$ and its y-intercept is -3, then the equation of the parabola is ________.

9 Given that the x-coordinates of the two points of intersection between the graph of a quadratic function and the x-axis are the two real roots to the equation $x^2 - 4x + 3 = 0$, and the y-coordinate of the turning point of the graph is $\sqrt{3}$, then the equation of the quadratic function is ________.

10 Given that there are two points of intersection between the graph of the quadratic function $y = (m - 1)x^2 + 2x + m$ and the two axes, then the value of m is ____________.

C. Questions that require solutions

11 The graph of a quadratic function $y = x^2 + bx + c$ passes through the points $(1, 3)$ and $(4, 3)$.

(a) Find the equation of the quadratic function and indicate the coordinates of the turning point of the graph and its y-intercept.

(b) After translating the graph of the function 1 unit in the negative y-direction, find the value of x for which the value of the resulting function is $y < 0$.

12 The line $y = x - 5$ and a parabola $y = -x^2 + bx + c$ intersect at two points, A and B. The x-coordinate of point A is -2, and point B is on the x-axis. C and D are two points on the line segment AB from left to right, and their x-coordinates are t and $t + 3$, respectively. Lines CF and DE are parallel to the y-axis and intersect the parabola at points F and E.

(a) Find the equation for the parabola.

(b) Is it possible that the quadrilateral $CDEF$ is a parallelogram? If so, find the value of t. If not, give a reason for your answer.

13 In the Cartesian coordinate system, the turning point A of a parabola $y = x^2 - 2mx + n + 1$ lies on the negative part of the x-axis. The parabola intersects the y-axis at point B, the x-coordinate of point C on the parabola is 1 and $AC = 3\sqrt{10}$.

(a) Find the equation of the parabola.

(b) If there exists a point D on the parabola such that the line DB passes through the first, second and fourth quadrants and the distance from the origin O to the line DB is $\frac{8\sqrt{5}}{5}$, find the coordinates of point D.

3.6 Graphs of other functions

Learning objective

Plot and interpret graphs in real-life contexts and find approximate solutions to problems involving distance, speed and acceleration.

A. Multiple choice questions

1. Sam is practising running 800 metres in a sports field. If his average speed and the time taken are recorded as x and y, respectively, then the graph that can possibly represent the relationship between y and x is ( ).

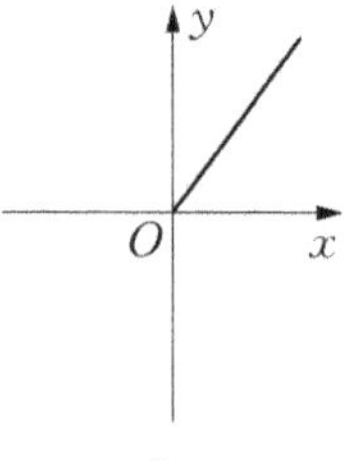

A.

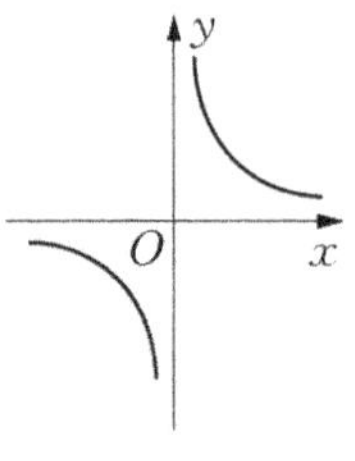

B.

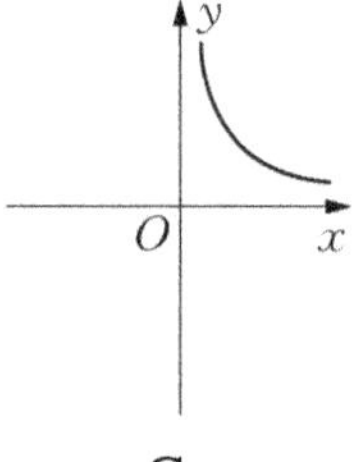

C.

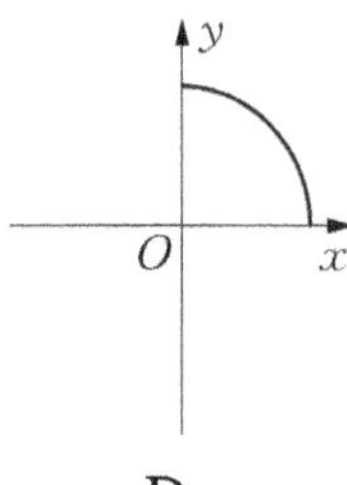

D.

2. The diagram shows a quadrilateral $ABCD$. A moving point P starts moving from point A along the route $A \to B \to C \to D$ at a constant speed, and stops when it reaches point D. During the movement, the area S of $\triangle PAD$ changes as the time t changes. Which graph may correctly represent such a relationship between S and t? ( )

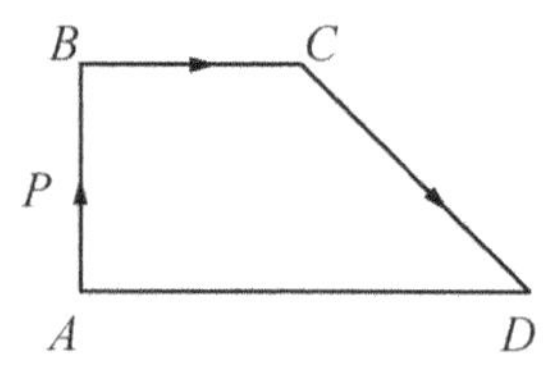

A.

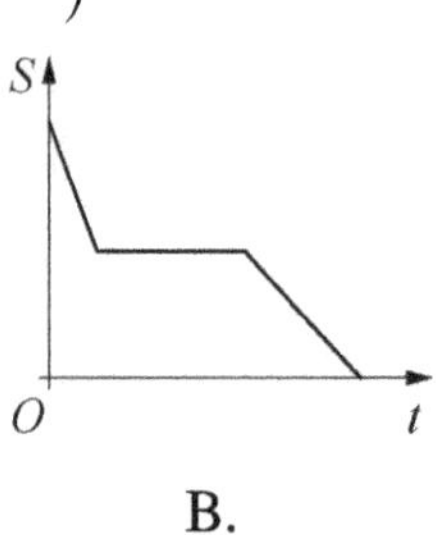

B.

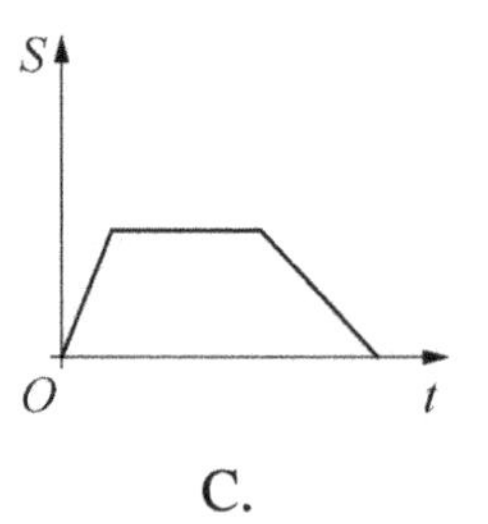

C.

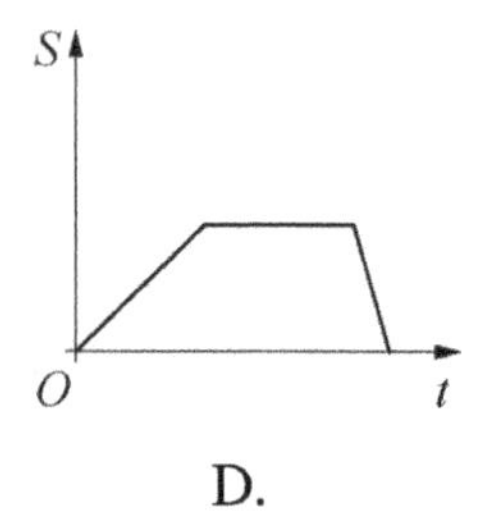

D.

3. The diagram shows a particle in motion and the relationship between its speed v (m/s) and the time t (seconds). The area of the shaded part represents ( ).

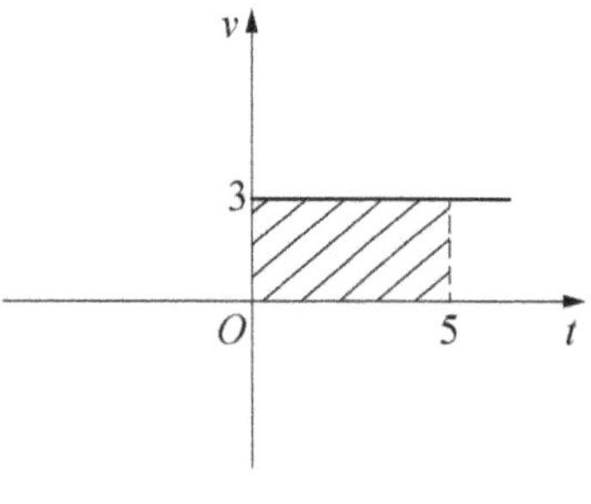

A. The distance of the motion of the particle when $t = 5$ seconds.

B. The speed of the motion of the particle when $t = 5$ seconds.

C. The distance of motion of the particle when $v = 3$ m/s.

D. The time of the motion of the particle when $v = 3$ m/s.

4 Ros walked from point A to point B along a straight road. The diagram shows the relationship between the distance s (metres) she walked and the time t (minutes) she took. Which of the following statements is correct? (   )

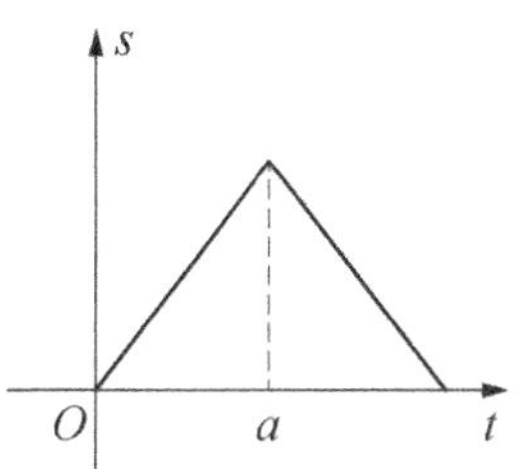

A. After Ros walked for a minutes towards point B at a constant speed, she continued walking towards point B at a slower speed.

B. After Ros walked for a minutes towards point B at a faster speed, she turned back towards point A at a slower speed.

C. After Ros walked for a minutes towards point B at a constant speed, she turned back towards point A at a constant speed.

D. After Ros walked for a minutes towards point B at a constant speed, she continued walking towards point B at a constant speed.

B. Fill in the blanks

5 The diagram shows the relationship between the cost of producing a product, y (in £1000), and the quantity of the product, x (in tonnes). To produce 10 tonnes of the product, the cost is ________ thousand pounds.

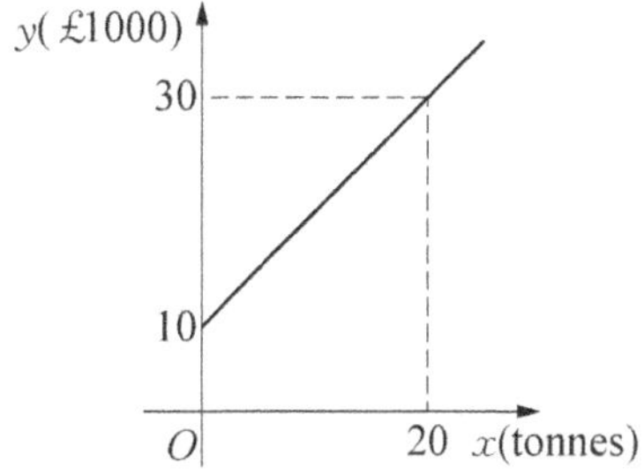

6 The relationship between the speed v of a particle M and the time t it has been travelling is shown in the diagram. The acceleration of this particle is ________.

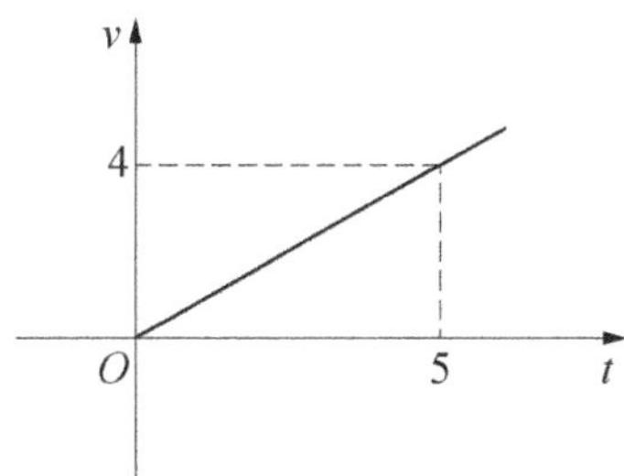

7 A particle N is accelerating at a rate of $2\,\text{m/s}^2$ in a straight line. The relationship between its speed v and the time t since it started moving is shown in the diagram. The initial speed of this particle is ________.

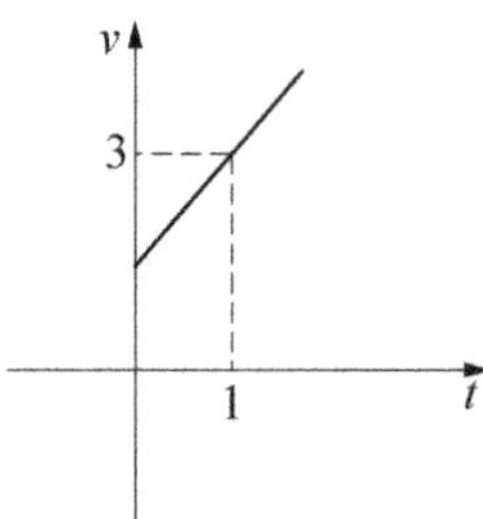

8 A certain kind of tax is charged at a progressive rate depending on the related expenditure. The diagram shows the relationship between the tax rate, $y\%$, and the amount of expenditure, x (in £1000). From the graph, for an expenditure of 140000 pounds, the amount of tax charged is ________ pounds.

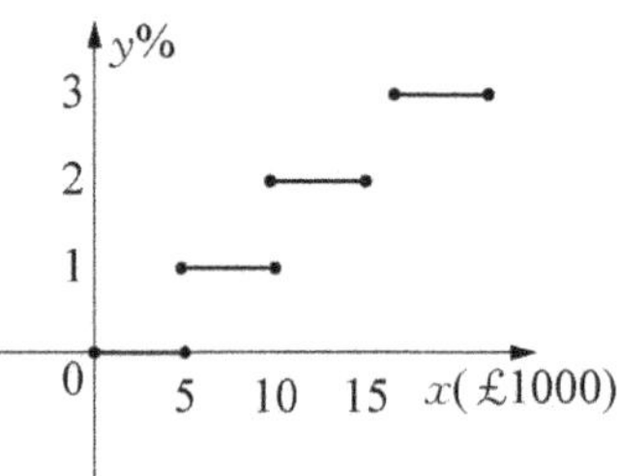

9 A ball falls freely from the top of a building 12 metres high (you can ignore air resistance). The graph representing the relationship between the distance of the ball above the ground h (metres) and the time t (seconds) taken to fall is part of a ________-shaped parabola, the equation of the line of symmetry is ________ and the y-intercept is ________.

10 Priya and Cathy start walking towards each other from two points A and B at the same time as shown by the lines l_1 and l_2, which intersect at point P. From the graph, the speed at which Priya walks is ________ km/h.

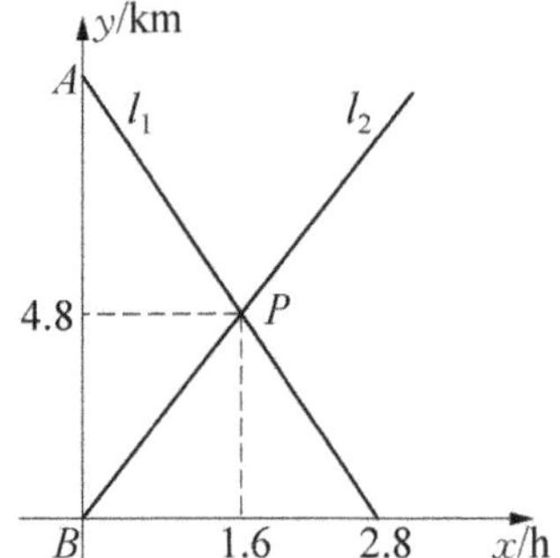

C. Questions that require solutions

11 Jim and his father set off from their house on a journey by scooter. Half an hour after leaving home, they arrived at town A where they stopped for a while before continuing to town B. 1.5 hours after they left home, Jim's mother set off by car from Jim's house and drove along the same route to town B. The diagram shows the graph of the distance y (kilometres) they travelled from home and the time t (hours) they took.

(a) Using the graph, find how long Jim and his father spent in town A.

(b) Find the speed at which Jim and his father travelled on each part of their journey and find the speed at which Jim's mother drove.

(c) Find the distance Jim and his father had travelled when his mother left home.

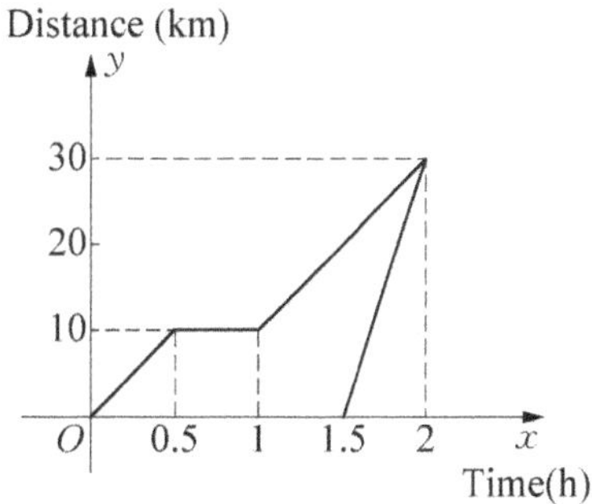

12 The diagram shows how the amount of a substance (unit: mg/100 ml) changed with the time (unit: hour) in a scientific experiment. It was found that for the first 1.5 hours (including 1.5 hours) the relationship between the amount of substance, y (mg/100 ml), and the time, x (hours), can approximately be represented by a quadratic function $y = -200x^2 + 400x$. Thereafter, the relationship between y and x can be approximately represented by the inverse proportional function $y = \frac{k}{x}$ $(k > 0)$.

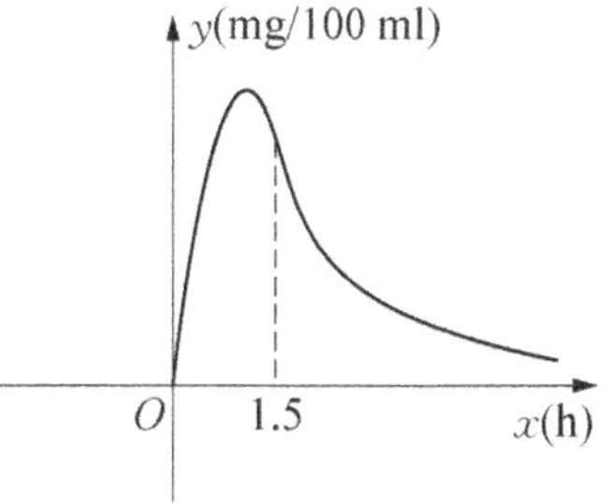

(a) How much time will it take for the amount, y, to reach its maximum? What is the maximum value?

(b) Find the value of k.

(c) The experiment starts at 7:00 a.m. Will the amount of the substance be above or below the 80 mg/100 ml level at 9:30 a.m.? Give a reason for your answer.

13 (a) A particle starts from rest and travels in a straight line at constant speed. The diagram shows the relationship between the speed v of a particle and the time t it has been travelling. Find the distance the particle has moved when $t = 3$.

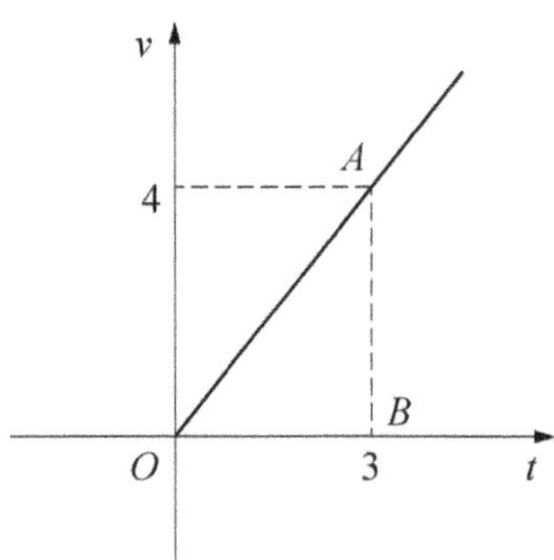

(b) The particle is moving in a straight line at constant speed from an initial speed v_0. The diagram shows the relationship between its speed v and the time t for which it has been moving. Explain how to use the graph to find the distance travelled in time t.

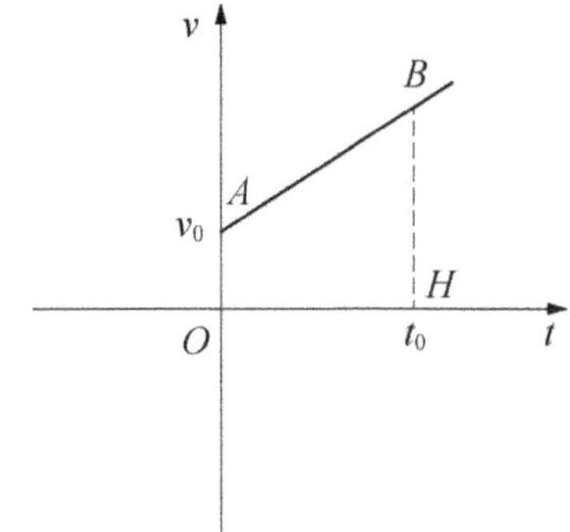

3.7 Gradient at a point on a curve

Learning objective

Calculate or estimate the gradients of graphs.

A. Multiple choice questions

1 Given that there are three points $A(x_1, y_1)$, $B(x_2, y_2)$ and $C(x_3, y_3)$, close to each other, on the graph of a function $y=f(x)$, then which of the following statements is incorrect? (   )

A. The gradient of the line AB is $\frac{y_2-y_1}{x_2-x_1}$, i. e. $\frac{\Delta y}{\Delta x}$.

B. When the independent variable, x, changes from x_1 to x_2, the average rate of change of the function $y=f(x)$ is $\frac{f(x_2)-f(x_1)}{x_2-x_1}$, i. e. $\frac{\Delta y}{\Delta x}$.

C. If $x_2-x_1=x_3-x_2$, then $\frac{y_2-y_1}{x_2-x_1}=\frac{y_3-y_2}{x_3-x_2}$.

D. If the graph of the function $y=f(x)$ is a straight line, and $x_2-x_1=x_3-x_2$, then $\frac{y_2-y_1}{x_2-x_1}=\frac{y_3-y_2}{x_3-x_2}$.

2 Given that the function $s=t^2+3$ shows the relationship between the distance S a particle moves and the time t it takes, then between the time period from $t_1=3$ to $t_2=3+m$ $(m>0)$, the average speed of the moving particle is (   ).

A. $6+m$ B. $6+m+\frac{9}{m}$ C. $3+m$ D. $9+m$

3 The average rate of change between any two points on each of the graphs of the three functions $y=-\frac{1}{2}$, $y=-\frac{1}{2}x$ and $y=-\frac{1}{2x}(x>0)$ is denoted by t_1, t_2 and t_3, respectively. Comparing the values of t_1, t_2 and t_3 gives (   ).

A. $t_1=t_2=t_3$ B. $t_1<t_2<t_3$

C. $t_2<t_1<t_3$ D. $t_3<t_2<t_1$

4 There are three points $A(x_1, y_1)$, $B(x_2, y_2)$ and $C(x_3, y_3)$ on the graph of the function $y = 2^x$. From left to right the gradients m_1, m_2 and m_3 are denoted by $m_1 = \frac{y_2 - y_1}{x_2 - x_1}$, $m_2 = \frac{y_3 - y_2}{x_3 - x_2}$ and $m_3 = \frac{y_3 - y_1}{x_3 - x_1}$. Comparing the values of m_1, m_2 and m_3 gives ( ).

A. $m_1 < m_2 < m_3$ B. $m_1 < m_3 < m_2$ C. $m_2 < m_1 < m_3$ D. $m_3 < m_1 < m_2$

B. Fill in the blanks

5 When $-3 \leqslant x \leqslant 1$, the average rate of change of the function $f(x) = 2x - 1$ is ________.

6 Given that the average rate of change between two points A and B on the graph of a proportional function is -3, then the equation of the proportional function is ________.

7 Given the inverse proportional function $f(x) = \frac{k}{x}(k \neq 0)$, and that when $x_1 < x_2 < 0$, $\frac{f(x_2) - f(x_1)}{x_2 - x_1} < 0$, then the set of values that k can take is ________.

8 Given that the average rate of change for any interval of two points on the graph of a function $y = f(x)$ is negative, then the value of the function y ________ as the independent variable x increases.

C. Questions that require solutions

9 A particle M is moving along a straight line with variable speed, which can be described using the function $s = 3t^2 + 2t$, where the distance s is in metres and the time t is in seconds. Find the average speeds from the 5th second to the 8th second and from the 8th second to the 11th second.

10 Find the average rate of change of the function $f(x) = \frac{2}{x}$ from $x = x_0$ to $x = x_0 + m$ $(m > 0)$.

11 $P(2, f(2))$ and $Q(2 + m, f(2 + m))$ are two points close to each other on the graph of the function $f(x) = -x^3$.

(a) Use a formula with m to express the gradient of the straight line PQ.

(b) Given the following definition: when point Q approaches point P until it coincides with P, we define the straight line QP as "the tangent line l to the curve (graph) at the point P", make an estimate of the gradient of the tangent line l, and write an equation for the function representing the tangent line l.

Unit test 3

A. Multiple choice questions

1 To obtain the graph of the quadratic function $C': y = 2 + \sqrt{x+1}$, the graph of $C: y = \sqrt{x}$ should be translated ( ).

A. 1 unit in the negative x-direction and then 2 units in the negative y-direction

B. 1 unit in the negative x-direction and then 2 units in the positive y-direction

C. 2 units in the negative x-direction and then 1 unit in the positive y-direction

D. 1 unit in the positive x-direction and then 2 units in the negative y-direction

2 Given that the graph of $y = x^2 + ax + 1$ is always above the x-axis no matter what value x takes, then the value of a is ( ).

A. $-1 < a < 1$ B. $-2 < a < 2$ C. $2 < a < \infty$ D. $-2 \leqslant a \leqslant 2$

3 For the quadratic function $y = x^2 - ax + 3 (a > 0)$, the line of symmetry of the graph is $x = m$. Therefore ( ).

A. $m > 1$ B. $m < 1$ C. $m > 0$ D. $m < 0$

4 Given a linear function $y = ax + b$ and a quadratic function $y = ax^2 - 2x - 2 (a \neq 0)$, their graphs in the same Cartesian coordinate system may be ( ).

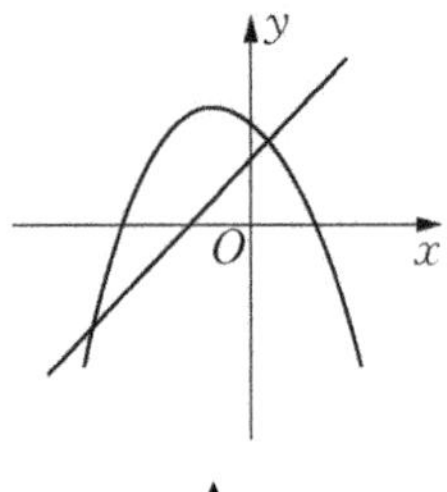

A.

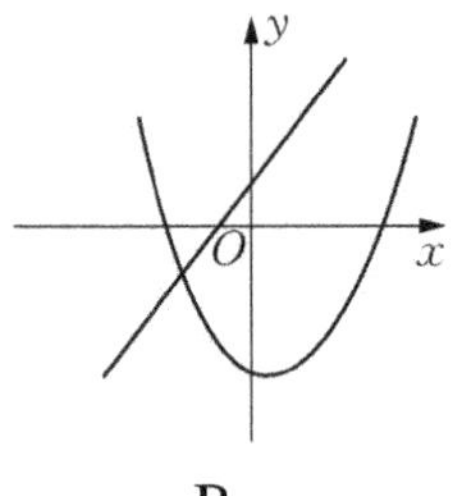

B.

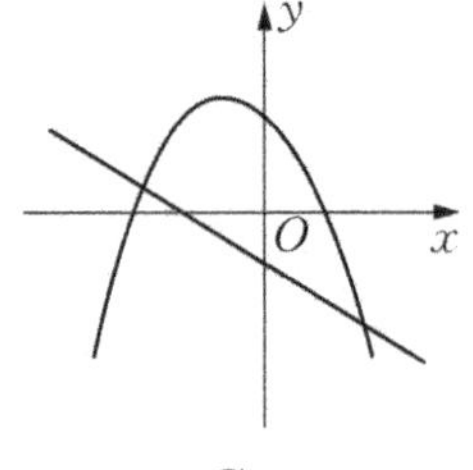

C.

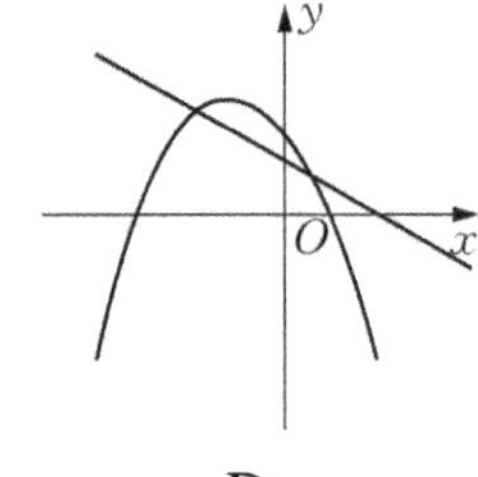

D.

5 The diagram shows part of a parabola $y = ax^2 + bx + c (a \neq 0)$. The point (ab, c) is in ( ).

A. the first quadrant

B. the second quadrant

C. the third quadrant

D. the fourth quadrant

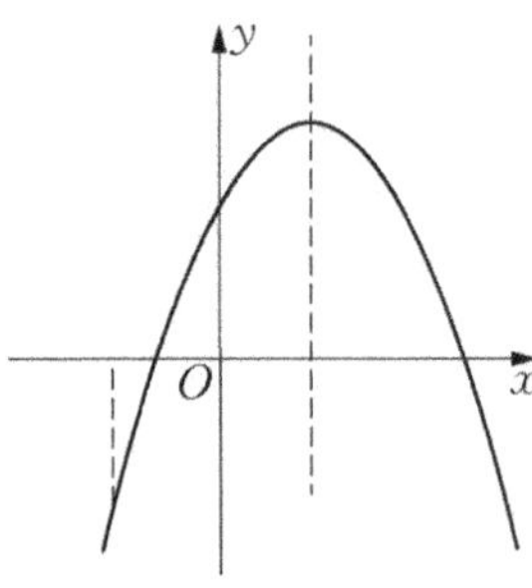

6 If a bottle of orange juice is poured into a glass at a constant speed, then the relationship between the height of juice, h, in the glass and the time, t, may be expressed as ( ).

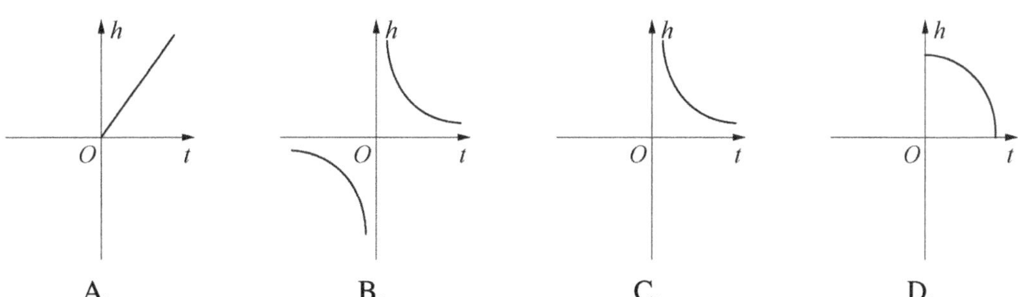

7 A tortoise and a hare were having a race. They started from the same spot at the same time. The hare ran faster and left the tortoise far behind, so the hare took a rest under a tree and fell asleep. The tortoise kept moving all the time, without stopping. Before the hare woke up, the tortoise was almost approaching the finishing line, so the hare got up and ran as fast as he could to catch up with the tortoise, but the tortoise won the race in the end. Which of the following graphs could show the changes in the distance S of the hare from the endpoint and the time t it took?

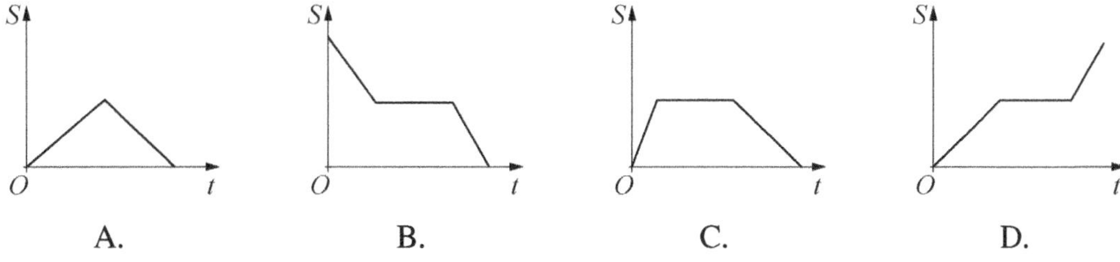

B. Fill in the blanks

8 The line $y = 3x - 5$ is reflected in the y-axis and then translated 2 units down. An equation of the resulting line is ________.

9 After reflecting the parabola $y = x^2 - 2x + 3$ in the y-axis, an equation of the resulting parabola is ________.

10 Write two properties of the function $y = x^2 - 2x + 4$.

(a) ________________________________

(b) ________________________________.

11 An equation for the line that passes through the point $A(-1, 3)$ and is parallel to the line l: $3x + 2y - 1 = 0$ is ________.

12 If there are two intersections between the line $y = kx + 2$ and the parabola $y = x^2 + 2x - 3$, then the set of values that k can take is ________.

13 The equation of the line of symmetry line of the quadratic function $y = -4x^2 + 3x - 2$ is ________.

14 The graph shows the quadratic function $y = ax^2 + bx + c$. The equation of the quadratic function is ________.

15 A quadratic function has the equation $y = -2x^2 - 3x + 7$. When ________, the y value of the function decreases when x increases, and when ________, the y value of the function increases when x increases.

16 A snail is climbing up a tree. The relationship between the speed v and the time t it takes is shown in the diagram. When $t = 2$, the distance the snail has climbed is ________.

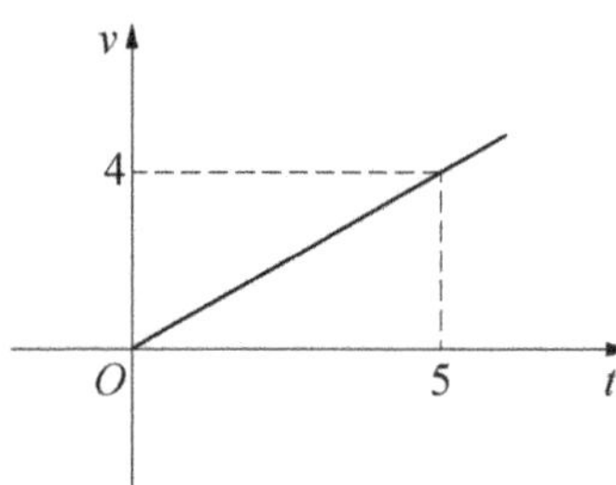

C. Questions that require solutions

17 Determine the number of solutions to the simultaneous equations $y = x^2 + 4x - 3$ and $y = 2x + a$ for different values of a.

18 The diagram shows part of a parabola $y = ax^2 + bx + c(a \neq 0)$.

(a) Complete the left-hand part of the diagram.

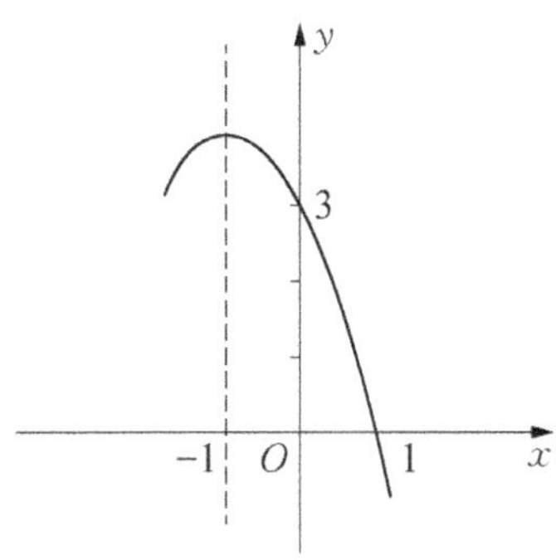

(b) Find an algebraic expression for the quadratic function.

19 A curve has the equation $y = x^2 - x + 3$ and a line has the equation $y = 3x - a$, where a is a constant.

(a) Show that the x-coordinates of the points of intersection of the line and the curve are given by the equation $x^2 - 4x + (3 + a) = 0$.

(b) The line intersects the curve at two points, and the x-coordinate of one of the points of intersection is −1. Find the x-coordinate of the other point of intersection.

(c) If the line $y = 3x - a$ is a tangent to the curve $y = x^2 - x + 3$ at point Q, then find the value of a at the point Q.

20 O is the origin, and A and B are two points on the graph of the function $y = ax^2 (a \geqslant 1)$. The x-coordinates of points A and B are -1 and 2, respectively, and AOB is a right-angled triangle.

(a) Find the value of a.

(b) What is the area of the triangle AOB?

21 $y = (m + 2)x^{m^2 + m - 4} + 8x - 1$ is a quadratic function.

(a) Find the possible values of m.

(b) Find the value of m for which the parabola for the quadratic function has a maximum point. Give the coordinates of the turning point.

(c) Find the interval for x in which the y value of the function decreases when x increases.

22 A diver jumps from high diving platform to water below. The relationship between the height h of the diver from the surface of the water and the time t after they leave the platform has the equation $h(t) = -4.9t^2 + 6.5t + 10$.

(a) Find the average speed of the diver between $t = 1$ and $t = 2$.

(b) If the diving platform is 10 metres above the surface of the water, how long does it take the diver to enter the water after leaving the platform?

(c) Sketch a graph of height above the water surface against time for the period between the diver leaving the platform and entering the water.

Chapter 4 Trigonometric ratios and trigonometric functions

4.1 Evaluating trigonometric ratios of acute angles (1)

Learning objective

Know and calculate with the exact values of $\sin\theta$ and $\cos\theta$ for $\theta = 0°, 30°, 45°, 60°$ and $90°$; and of $\tan\theta$ for $\theta = 0°, 30°, 45°$ and $60°$.

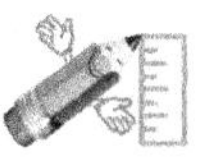

A. Multiple choice questions

1 The value of $\sqrt{2}\sin 45°$ is ( ).

A. $\sqrt{2}$ B. 1 C. $\frac{\sqrt{2}}{2}$ D. $\frac{1}{2}$

2 Which of the following equations is correct? ( )

A. $\cos\frac{\sqrt{2}}{2} = 45°$ B. $\cos 45° = 1$ C. $\frac{1}{\sqrt{3}} = \tan 30°$ D. $\sin\frac{1}{2} = 30°$

3 In right-angled triangle ABC, $\angle C = 90°$ and the sides opposite angles A, B and C are a, b and c respectively. Which of the following equations is not necessarily true? ( )

A. $b = a\tan B$ B. $a = c\cos B$ C. $c = \frac{a}{\sin A}$ D. $a = b\cos A$

4 In right-angled triangle ABC, $\angle ACB = 90°$. If side $AB = 2$ units and $\tan CAB = \frac{1}{2}$, then side AC is ( ).

A. $\frac{\sqrt{5}}{5}$ B. $\frac{2\sqrt{5}}{5}$ C. $\frac{3\sqrt{5}}{5}$ D. $\frac{4\sqrt{5}}{5}$

B. Fill in the blanks

5 Calculate: $2\sin 60° + \tan 45° =$ ________.

6 If angle A is an acute angle and $\tan A = \frac{\sqrt{3}}{3}$, then $\cos A =$ ________.

7 Given that A is an acute angle and $\sin A = \frac{5}{13}$, then $\tan A =$ ________.

8 In right-angled triangle ABC, $\angle ACB = 90°$. If $BC = \sqrt{6}$ and $\angle ABC = 60°$, then side AB is ________.

9 In isosceles triangle ABC, if $AB = AC = 5$ and $BC = 6$, then the cosine of the base angle equals ________.

C. Questions that require solutions

10 Calculate:

(a) $\sin^2 45° \times \cos 60° + \tan 45° \times \cos^2 30°$

(b) $\dfrac{\sin^2 30° + \sin^2 60°}{\tan 60° \times \dfrac{1}{\tan 30°}} + \cos^2 45°$.

11 In acute-angled triangle ABC, if $\sin A = \frac{\sqrt{3}}{2}$ and $\angle B = 75°$, find the value of $\cos C$.

12 The diagram shows triangle ABC. $\angle ACB = 90°$. AB and CD are perpendicular and intersect at point D. If $\angle ABC = 30°$ and $CD = 6$ units, find the length of AB.

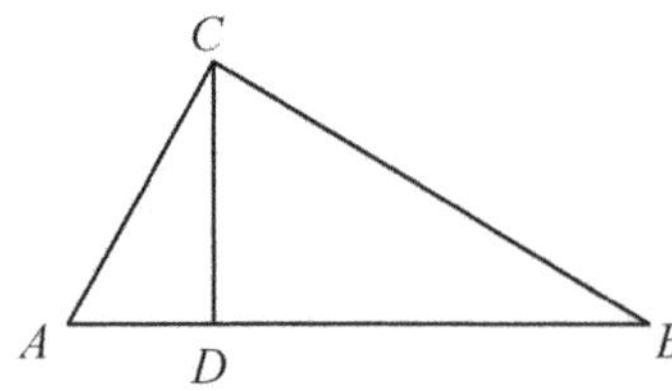

4.2 Evaluating trigonometric ratios of acute angles (2)

Learning objective

Know and calculate with the exact values of $\sin\theta$ and $\cos\theta$ for $\theta = 0°$, $30°$, $45°$, $60°$ and $90°$; and of $\tan\theta$ for $\theta = 0°$, $30°$, $45°$ and $60°$.

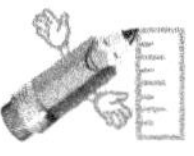

A. Multiple choice questions

1 In triangle ABC, if both $\angle BAC$ and $\angle ABC$ are acute and $\sin BAC = \cos ABC = \frac{1}{2}$, then the triangle ABC is ( ).

A. an obtuse triangle
B. a right-angled triangle
C. an acute-angled triangle
D. not possible to tell

2 In right-angled triangle ABC, if the length of each side is tripled, then the values of the three trigonometric ratios of the acute angle A are ( ).

A. all tripled
B. all reduced by two thirds
C. unchanged
D. not possible to tell

3 If $\sin A = \frac{2}{3}$, then the range of the possible value of acute angle BAC is ( ).

A. $0° < \angle A < 30°$
B. $30° < \angle A < 45°$
C. $45° < \angle A < 60°$
D. $60° < \angle A < 90°$

4 If α is an acute angle, then which of following expressions is correct? ( )

A. $\sin\alpha + \cos\alpha = 1$
B. $\sin\alpha + \cos\alpha > 1$
C. $\sin\alpha + \cos\alpha < 1$
D. $\sin\alpha + \cos\alpha \leqslant 1$

B. Fill in the blanks

5 In acute-angled triangle ABC, if $\sin A = \frac{\sqrt{3}}{2}$ and $\cos B = \frac{\sqrt{2}}{2}$, then $\angle C =$ ________.

6 If an acute angle α satisfies the equation $\sqrt{2}\cos(\alpha + 30°) = 1$, then the angle α equals ________.

7 In triangle ABC, $\angle ACB = 90°$. If side $BC = \sqrt{15}$ and side $AC = \sqrt{5}$, then $\angle ABC =$ ________.

8 In a right-angled triangle ABC, $\angle C = 90°$. If $b : c = 1 : 3$, $\tan B =$ ________.

9 Given that the perimeter of an isosceles triangle is 20, if the cosine of one of the interior angles is $\frac{2}{3}$, then the length of the two equal sides of the isosceles triangle is ________.

C. Questions that require solutions

10 In a triangle ABC, $\angle A = 30°$. If $\tan B = \frac{\sqrt{3}}{3}$ and $AC = 2\sqrt{3}$, find the length of side AB.

11 The diagram shows a right-angled triangle ABC in which $\angle C = 90°$, $\angle BAC = 30°$ and $AD = AB$. Find the size of $\angle D$ and hence write the exact expressions of $\sin D$, $\cos D$ and $\tan D$, simplifying the surds.

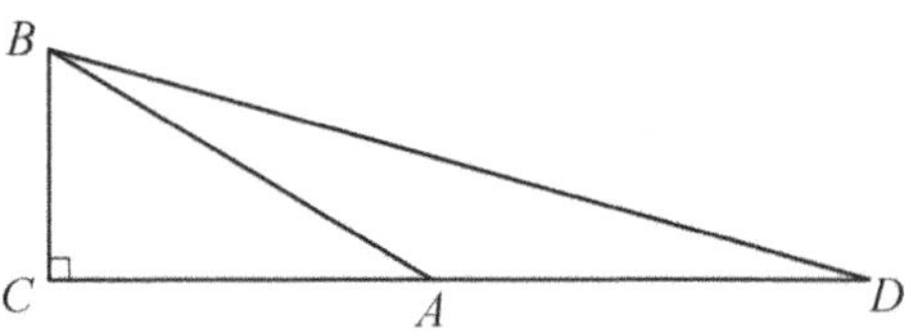

12 The diagram shows a rhombus $ABCD$. AE is perpendicular to BC. The length of the diagonal BD is 4 cm and $\tan CBD = \frac{1}{2}$.

(a) What is the length of side AB?

(b) What is the value of the sine of angle ABE?

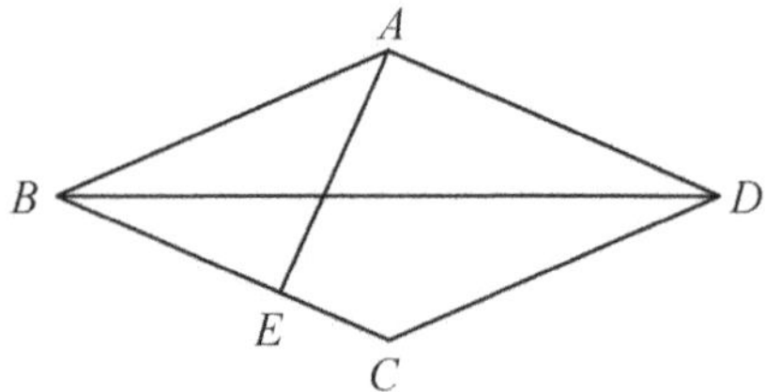

4.3 Angles of any size and their measures (1)

Learning objective

Know exact values for the sine, cosine and tangent of acute angles and use this to generate values for obtuse and reflex angles in all four quadrants.

A. Multiple choice questions

1 If an angle can be written as $k \times 360° + \alpha (k \in \mathbf{Z})$ when considering the rotation of a line of unit length in a Cartesian coordinate plane, then angle α (　　).

A. must be less than 90°　　B. must be in the first quadrant

C. must be a positive angle　　D. can be any angle

2 Which of these is equivalent to an angle of −45° when considering the rotation of a line of unit length in a Cartesian coordinate plane? (　　)

A. $k \times 360° + 45° (k \in \mathbf{N})$　　B. $k \times 360° + 45° (k \in \mathbf{Z})$

C. $k \times 360° - 45° (k \in \mathbf{N})$　　D. $k \times 360° - 45° (k \in \mathbf{Z})$

3 Which of the following four statements are equivalent? (　　)

① θ is an acute angle

② θ is not an obtuse angle

③ θ is neither an obtuse angle nor a right angle

④ $0° < \theta < 90°$

A. ① and ②　　B. ② and ③　　C. ③ and ④　　D. ① and ④

4 Which of the following statements about angles in a Cartesian coordinate plane is correct? (　　)

A. Any angle whose terminal side in the first quadrant must be an acute angle.

B. Angles sharing the same terminal sides must be equal.

C. Equal angles must share the same initial and terminal sides.

D. Unequal angles must have different terminal sides.

The **initial side** of an angle is the position of the first line that forms the angle, and the **terminal side** is the position of this line after it is rotated.

B. Fill in the blanks

5 In a Cartesian coordinate plane, when considering the rotation of a line of unit length, any angle with the same terminal side as $120°$ can be written as ________.

6 If the initial line is on the positive x-axis of a Cartesian coordinate plane, the terminal side of angle $225°$ lies in the ________ quadrant.

7 If the initial line is on the positive x-axis of a Cartesian coordinate plane, the terminal side of angle $-20°$ lies in the ________ quadrant.

8 If the initial line is on the positive x-axis of a Cartesian coordinate plane, a positive angle less than $360°$ with the same terminal side as angle $-45°$, is ________.

9 If the initial line is on the positive x-axis of a Cartesian coordinate plane, write down all angles whose terminal sides coincide with the:

(a) positive side of the x-axis ________________

(b) negative side of the x-axis ________________

(c) positive side of the y-axis ________________

(d) negative side of the y-axis ________________.

C. Questions that require solutions

10 If the initial line β is on the positive x-axis of a Cartesian coordinate plane, find the angle β whose terminal side coincides with that of the angle α and $0° \leqslant \beta \leqslant 360°$, and determine which quadrant it is in.

(a) $\alpha = 800°$

(b) $\alpha = -400°$

11 Given that, in a Cartesian coordinate plane, angles α and β share the same terminal side, find the value of angle α in each question below.

(a) $-360° < \alpha \leqslant 0°$, $\beta = 15°$

(b) $360° \leqslant \alpha < 720°$, $\beta = -120°$

(c) $-720° \leqslant \alpha < -360°$, $\beta = 180°$

(d) $0° \leqslant \alpha < 360°$, $\beta = 400°$

12 Write down an angle whose terminal side is on the bisectors of the first quadrant and the third quadrant.

4.4 Angles of any size and their measures (2)

Learning objective

Know exact values for the sine, cosine and tangent of acute angles and use this to generate values for obtuse and reflex angles in all four quadrants, finding all possible values in a given range.

A. Multiple choice questions

1 If $\alpha = 45° + k \times 180°$, $k \in \mathbf{Z}$, then the terminal side of α lies in the ( ).

A. first quadrant or third quadrant
B. first quadrant or second quadrant
C. second quadrant or fourth quadrant
D. third quadrant or fourth quadrant

2 If the terminal sides of angles α and β are symmetrical with respect to the y-axis, then ( ).

A. $\alpha = 90° - \beta$
B. $\alpha = k \times 360° + 90° - \beta (k \in \mathbf{Z})$
C. $\alpha = 360° - \beta$
D. $\alpha = k \times 360° + 180° - \beta (k \in \mathbf{Z})$

B. Fill in the blanks

3 When the minute hand on a clock face rotates for 15 minutes, the angle it turns through is ________.

4 Express the following angles in form of $k \times 360° + \alpha (0 \leqslant \alpha < 360°, k \in \mathbf{Z})$.
$-108° =$ ________________; $-225° =$ ________________.

5 Given that the terminal side of angle α coincides with that of angle $20°$ and $0° < \alpha < 720°$, then there are ________ possible values of α that satisfy the conditions.

6 The angle whose terminal side coincides with that of angle $950°$ is in the ________ quadrant. The smallest positive of such angle is ________ and the largest negative of such angle is ________.

7 If α is an acute angle, the terminal side of $k \times 180° - \alpha (k \in \mathbf{Z})$ lies in the________ quadrant.

8 Given that $0° < \alpha < 360°$, and the terminal side of the angle which is 7 times angle α coincides with the terminal side of α, then there are at most ________ possible values that α can take.

9 Write down all the angles whose terminal sides coincide with:

(a) the x-axis: ________________.

(b) the y-axis: ________________.

(c) one of the coordinate axes: ________________.

10 Which of the angles 15°, 180°, −100°, 270°, 120°, −360°, −200°, 0°, 1200°, −180° and 460°, satisfy the following conditions:

(a) Is a positive angle: ________________.

(b) Is a negative angle: ________________.

(c) Has terminal side on the x-axis: ________________.

(d) Has terminal side in the second quadrant: ________________.

C. Questions that require solutions

11 The terminal side of angle α lies in the fourth quadrant.

(a) State the quadrant in which the terminal side of $\frac{\alpha}{2}$ lies.

(b) Write down the range of $\frac{\alpha}{3}$.

(c) Write down the range of 2α.

12 Given that angles α and β are both acute angles, $\alpha + \beta$ and $-280°$ have the same terminal side, and the terminal side of $\alpha - \beta$ coincides with that of $-670°$, find the value of α and the value of β.

4.5 Sine rule, cosine rule and solving scalene triangles (1)

Learning objective

Solve triangle problems using the sine and cosine rules.

A. Multiple choice questions

1 The diagram shows a right-angled isosceles triangle ABC, in which $\angle C = 90°$, $AC = 6$ and D is a point on AC. If $\tan \angle DBA = \frac{1}{5}$, then the length of AD is (　　).

A. $\sqrt{2}$　　　B. 2

C. 1　　　D. $2\sqrt{2}$

2 The diagram shows a right-angled triangle ABC in which $\angle C = 90°$ and D is a point on BC. If $\angle DAC = 30°$, $BD = 2$ and $AB = 2\sqrt{3}$, then the length of AC is (　　).

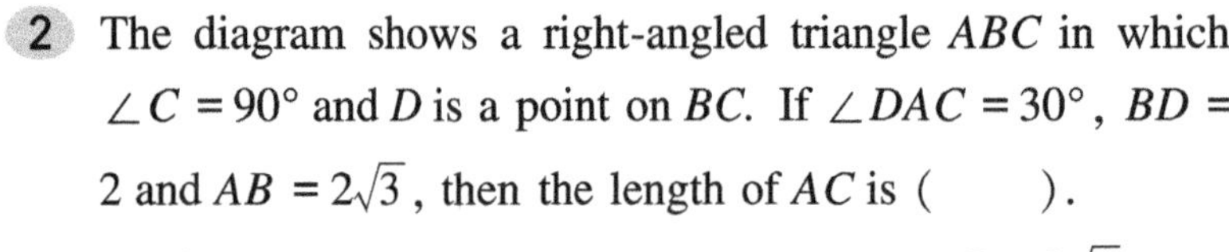

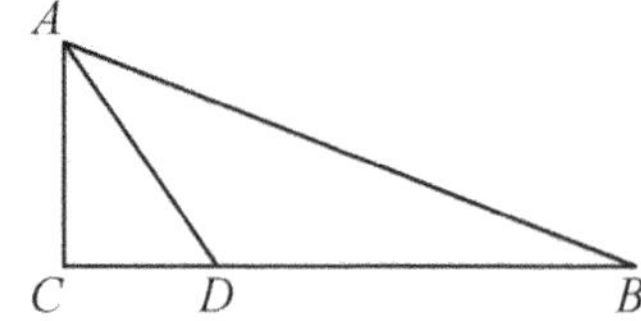

A. 3　　　B. $2\sqrt{2}$

C. $\sqrt{3}$　　　D. $\frac{3}{2}\sqrt{3}$

3 Out of the following pairs of conditions for a right-angled triangle, which one is not sufficient to find the lengths of all the sides and the sizes of all the angles? (　　)

A. given side a and angle A　　　B. given side c and angle B

C. given angle A and angle B　　　D. given side a and side b

B. Fill in the blanks

4 In a right-angled triangle with an angle 60°, the ratio of the side opposite to the 60° angle and the hypotenuse is ________.

5 In a right-angled triangle ABC, if $\angle C = 90°$, $a = 5$ cm and $b = 12$ cm, then $c =$ ________.

6 In a right-angled triangle ABC, if $\angle A = 90°$, $BC = 5$ cm and $\angle B = \alpha$, then $AB =$ ________. (Express the answer as a trigonometric ratio in terms of α.)

7 If α is one of the acute angles of a right-angled isosceles triangle, then the value of $\sin\alpha$ is ( ).

A. $\frac{1}{2}$ B. $\frac{\sqrt{2}}{2}$ C. $\frac{\sqrt{3}}{2}$ D. 1

8 If α is one of the acute angles of a right-angled isosceles triangle, then the value of $\cos\alpha$ is ( ).

A. 1 B. $\frac{1}{2}$ C. $\frac{\sqrt{3}}{2}$ D. $\frac{\sqrt{2}}{2}$

C. Questions that require solutions

9 In a right-angled triangle ABC, $\angle C = 90°$ and $BC : AC = 3 : 4$. Find the sine, cosine, and tangent of $\angle B$.

10 In the triangle ABC shown in the diagram, $AB = \frac{3}{2}$, $AC = 4$ and $\angle A = 60°$. Find the area of triangle ABC.

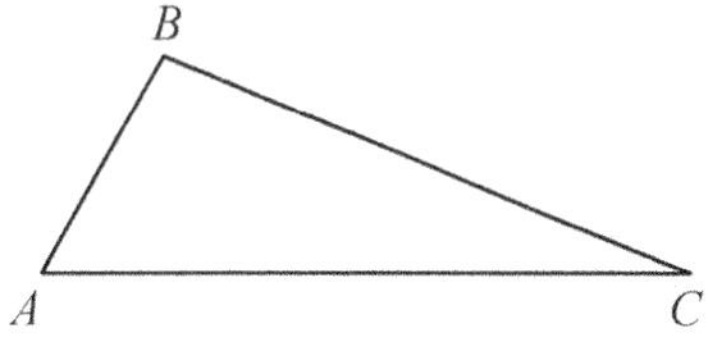

11 In a triangle ABC, $\angle C = 90°$, $3a = \sqrt{3}b$ and $c = 10$ cm. Find the unknown sides and angles of this right-angled triangle.

4.6 Sine rule, cosine rule and solving scalene triangles (2)

Learning objective

Solve triangle problems using the sine and cosine rules and the trigonometric formula for the area of any triangle.

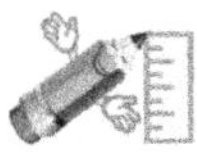

A. Multiple choice questions

1 To find two unknown sides in a right-angled triangle, you must be given ( ).

A. an acute angle

B. two acute angles

C. one side and one angle

D. two sides

2 In triangle ABC, $AB = AC = m$ and $\angle B = \alpha$. Then the length of BC is ( ).

A. $2m\sin\alpha$

B. $2m\cos\alpha$

C. $2m\tan\alpha$

D. $\dfrac{2m}{\tan\alpha}$

3 The diagram shows a rectangle $ABCD$. DE is perpendicular to AC and the lines intersect at point E. Let $\angle ADE = \alpha$, $\cos\alpha = \dfrac{3}{5}$ and $AB = 4$ cm. Then the length of AD is ( ).

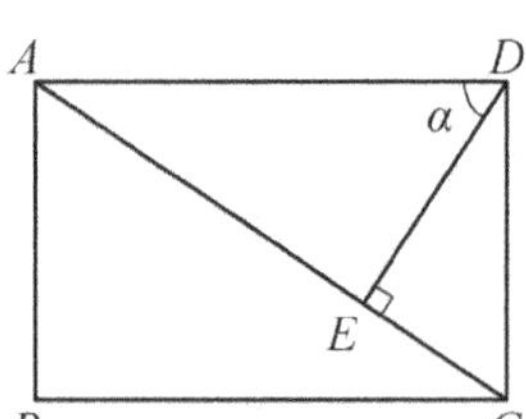

A. 3 cm

B. $\dfrac{20}{3}$ cm

C. $\dfrac{16}{3}$ cm

D. $\dfrac{16}{5}$ cm

4 A rectangular piece of paper $ABCD$ is folded about DE, as shown in the diagram and point C' is the image of the vertex C. $AB = 4$ cm. If $\angle C'ED = 30°$, then the length of the fold line ED is ( ).

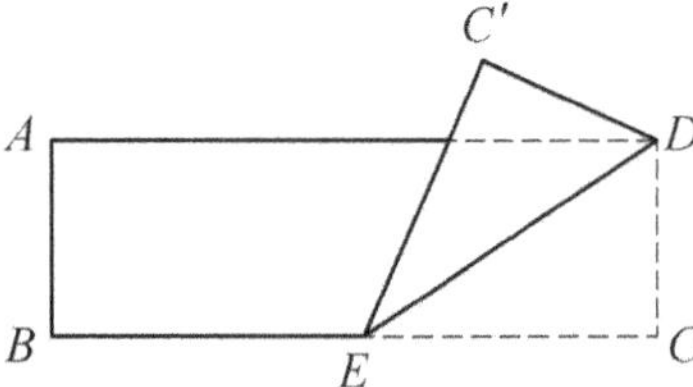

A. 4

B. $4\sqrt{3}$

C. 8

D. $5\sqrt{3}$

B. Fill in the blanks

5 In a right-angled triangle ABC, $\angle C = 90°$ and point D is the midpoint of AB. If $BC = 3\text{ cm}$ and $CD = 2\text{ cm}$, then $\cos DCB =$ ________.

6 The side length of a square $ABCD$ is 1. Line segment BD is rotated about point B, and the image point of D, which is point D', is on the extension of BC beyond C. Then $\tan \angle BAD' =$ ________.

7 In the diagram, $\tan AOB = \frac{4}{3}$, point P is on side OA and $OP = 5\text{ cm}$. M and N are on the side OB and $PM = PN$. If $MN = 2\text{ cm}$, then $PM =$ ________.

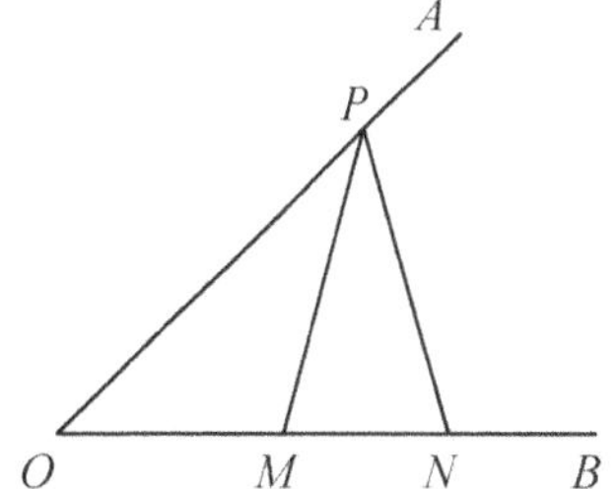

8 In a right-angled triangle ABC, $\angle ACB = 90°$, $AC = \sqrt{2}$ and $\cos A = \frac{\sqrt{3}}{2}$. If the triangle ABC is rotated about point C to obtain $\triangle A'B'C$, so that point B' is on the bisector of $\angle ACB$, and $A'B'$ intersects AC at point H, then the length of CH is ________.

C. Questions that require solutions

9 In a triangle ABC, $\angle C = 90°$, $\sin B = \frac{5}{13}$, point D is on side BC, $\tan DAC = \frac{3}{5}$ and $AB = 13\text{ cm}$. Find the length of CD.

10 In a right-angled triangle ABC, $\angle C = 90°$ and $AC = 12\,\text{cm}$. AD is the bisector of $\angle A$, and $AD = 8\sqrt{3}$. Find the lengths of BC and AB.

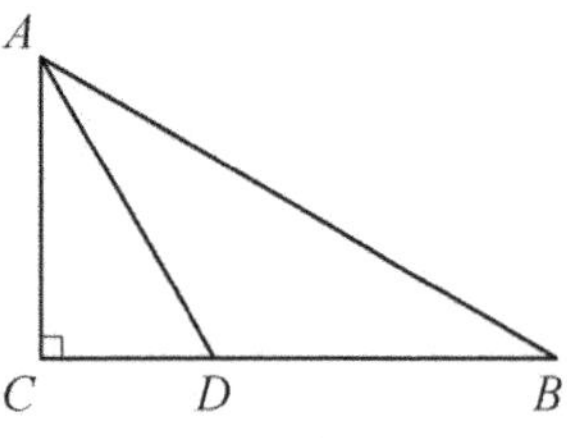

11 The diagram shows a triangle ABC, in which $BC = 9$, $AB = 6\sqrt{2}$ and $\angle ABC = 45°$.

(a) Find the area of $\triangle ABC$.

(b) Find the value of $\cos \angle C$.

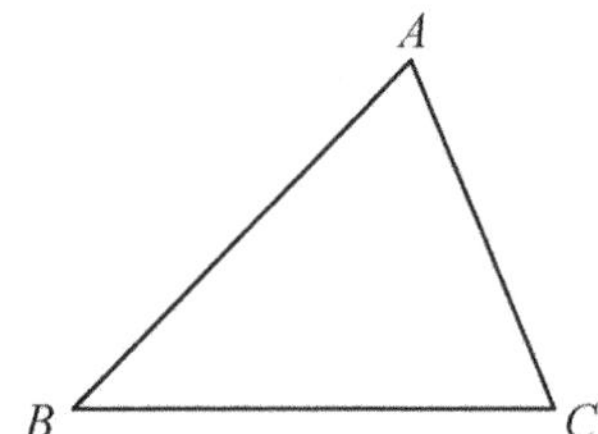

4.7 Sine rule, cosine rule and solving scalene triangles (3)

Learning objective

Solve problems in practical contexts involving the sine and cosine rules and the trigonometric formula for the area of any triangle.

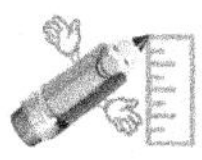

A. Multiple choice questions

1 A slope has an inclination of 30°. If Emmy walks down the ramp for 2 metres, then she goes down vertically by (   ).

A. 1 metre B. $\sqrt{3}$ metres C. $2\sqrt{3}$ metres D. $\frac{2\sqrt{3}}{3}$ metres

2 As shown in the diagram, the gradient of the slope formed by the conveyor belt and the ground is 1 : 2. If an object is sent from point A on the ground to point B 2 metres above the ground, then the distance the object travels from point A to point B is (   ).

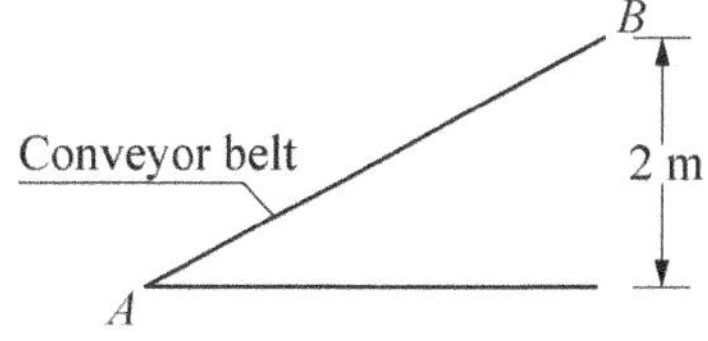

A. 4 metres B. $2\sqrt{3}$ metres C. $\sqrt{5}$ metres D. $2\sqrt{5}$ metres

3 The diagram shows points A and B, on different sides of a river, and point C, which is perpendicular to AB. The distance $AC = a$ and angle $ACB = \alpha$. Then $AB =$ (   ).

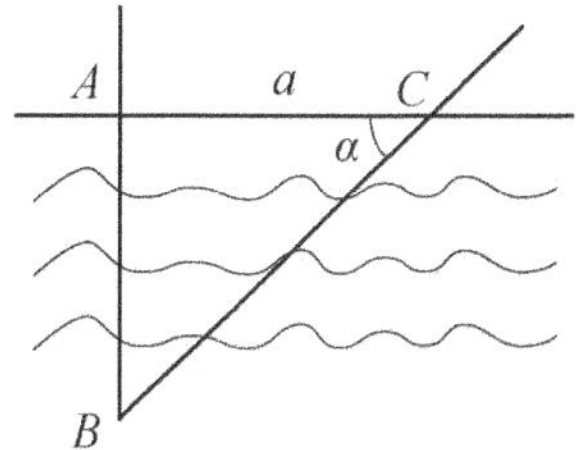

A. $a \sin\alpha$ B. $a \tan\alpha$

C. $a \cos\alpha$ D. $\frac{a}{\tan\alpha}$

4 AC is a guy wire used to provide stability to the vertical pole AB, as shown in the diagram. $BC = 6$ m and $\angle ACB = 52°$. The length of AC is (   ).

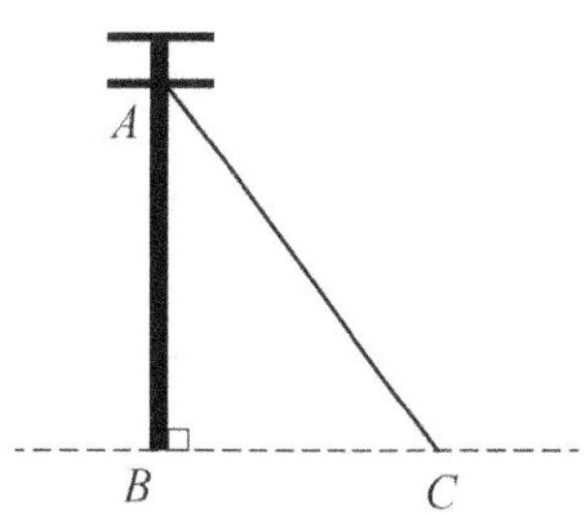

A. $\frac{6}{\sin 52°}$ metres B. $\frac{6}{\tan 52°}$ metres

C. $6\cos 52°$ metres D. $\frac{6}{\cos 52°}$ metres

B. Fill in the blanks

5 In an isosceles triangle ABC, if $AB = AC = 3$ and $BC = 2$, then the value of sine of the base angle is ________.

6 In a parallelogram $ABCD$, the lengths of the two adjacent sides are 4 cm and 6 cm, and their included angle is 60°. The length of the shorter diagonal is ________.

7 The diagram shows a right-angled triangle ABC with $\angle C = 90°$, $AC =$ 4 cm and $BC = 3$ cm. The triangle is rotated 90° clockwise about point C. The new positions of vertices A and B are D and E, respectively. The value of $\tan \angle ADE =$ ________.

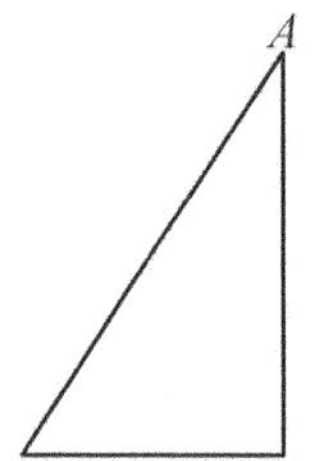

8 In the diagram shown, $AD = 6$ cm and $\sin ABC = \sin BCD = \frac{3}{5}$. If the height of this isosceles trapezium is 9 cm, then $BC =$ ________ cm.

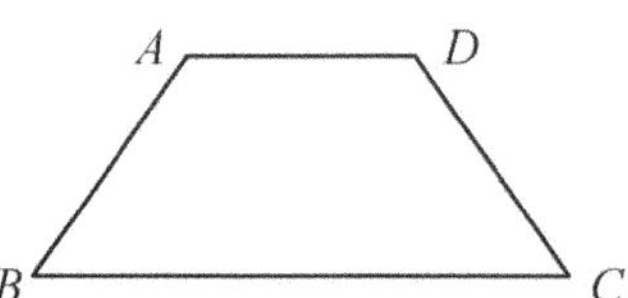

9 In a triangle ABC, $\angle C = 90°$, $AB = 9$ cm and $\cos A = \frac{2}{3}$. The triangle is rotated clockwise about point C so that the images of points A and B are A' and B', respectively. If point A' is on side AB, then the distance between points B and B' is ________.

C. Questions that require solutions

10 In a rectangle $ABCD$, $AB = 4$ cm, $BC = 5$ cm and E is a point on side AB. Triangle BCE is folded along CE so that point F is the image of point B, which is on side AD. Find the value of tangent of $\angle AFE$.

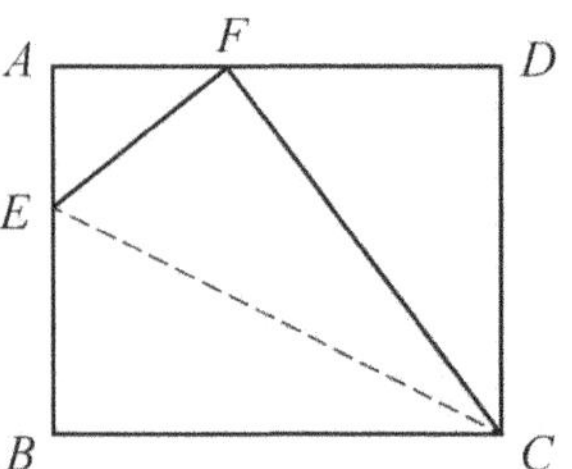

11 The diagram shows a triangle ABC in which $\angle A = 30°$, $\angle B = 45°$ and $AC = 8\,\text{cm}$. Point P is on line segment AB. A straight line is drawn from C to P, and $\tan \angle APC = \frac{4}{3}$.

(a) Find the length of CP.

(b) Find the value of $\sin BCP$.

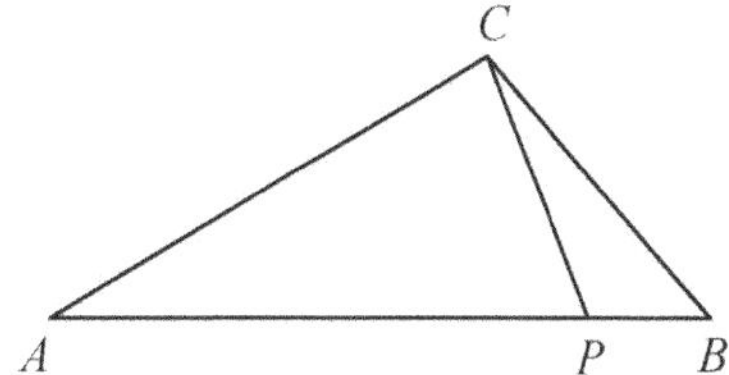

4.8 Sine rule, cosine rule and solving scalene triangles (4)①

Learning objective

Solve problems involving any triangle using the sine and cosine rules and the trigonometric formula for the area of any triangle.

A. Multiple choice questions

1 In the triangle ABC, $\angle B = 60°$ and $b^2 = ac$. Then triangle ABC must be (　　).

A. a right-angled triangle　　B. an obtuse-angled triangle

C. an isosceles triangle　　D. an equilateral triangle

2 In a triangle ABC, if $2\cos B \sin A = \sin C$, then triangle ABC must be (　　).

A. a right-angled isosceles triangle　　B. a right-angled triangle

C. an equilateral triangle　　D. an isosceles triangle

B. Fill in the blanks

3 If the ratio of the three interior angles of a triangle is 1 : 2 : 3, then the ratio of their three corresponding sides is ________.

4 In a triangle ABC, if $a = 7$, $b = 7\sqrt{3}$ and $\angle B = 60°$, then $\angle A =$ ________.

5 In a triangle ABC, if $a = 4$, $c = 3$ and $\angle C = 45°$, then $\sin A =$ ________.

6 In a triangle ABC, if $a = 6$, $b = 8$ and $\cos C = -\frac{1}{2}$, then $c =$ ________.

7 In a triangle ABC, if $a = 10$, $b = 24$ and $c = 26$, then $\angle C =$ ________.

8 In a triangle ABC, $\angle A = 60°$, $AB = 5$ and $BC = 7$. Therefore $AC =$ ________.

---

① In this unit, angles $\angle A$, $\angle B$ and $\angle C$ of $\triangle ABC$ are opposite to sides a, b and c, respectively.

9 In a triangle ABC, $AB = \frac{3}{2}$, $AC = 4$ and $\angle A = 60°$. Then, the area of triangle ABC = ________.

10 In a triangle ABC, if $\angle A = 60°$, $b = 12$ and the area of triangle $ABC = 24\sqrt{3}$, then c = ________.

C. Questions that require solutions

11 In a triangle ABC, $a\cos A = b\cos B$. Determine what type of triangle this is.

12 In a triangle ABC, given that $a = 4, b = 5$ and the area of triangle $ABC = 5\sqrt{3}$, find c.

13 In a triangle ABC, given that $a = 2\sqrt{2}$, $b = 2\sqrt{3}$, $\angle A = 45°$ and $b < c$, find the length of the third side c, and the sizes of angles B and C.

4.9 Trigonometric functions (1)

Learning objective

Recognise, sketch and interpret graphs of the trigonometric functions $y = \sin x$, $y = \cos x$, and $y = \tan x$ for angles of any size.

A. Multiple choice questions

1 Which of these is the graph of the function $y = \sin x$ $(-180° \leqslant x \leqslant 180°)$? (   )

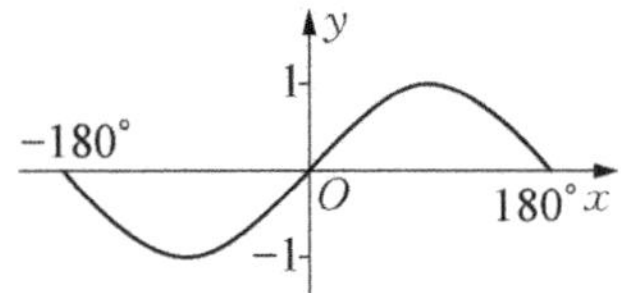

A.

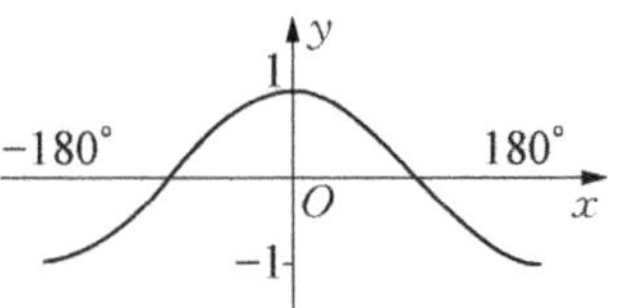

B.

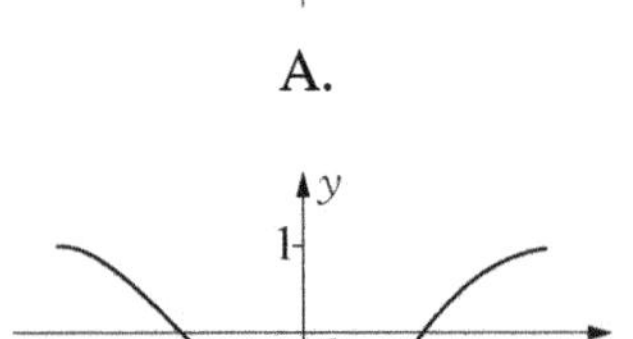

C.

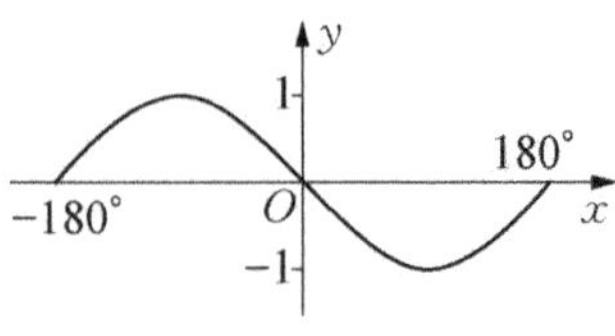

D.

2 Which of these is the graph of the function $y = -\sin x$ $(-360° \leqslant x \leqslant 360°)$? (   )

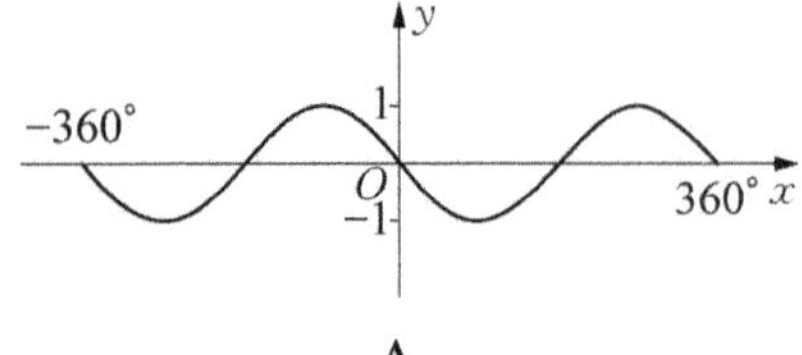

A.

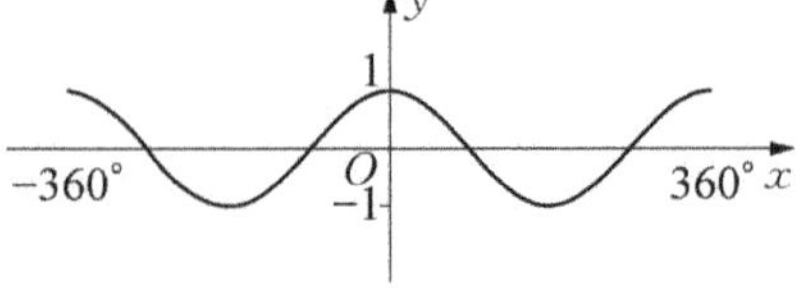

B.

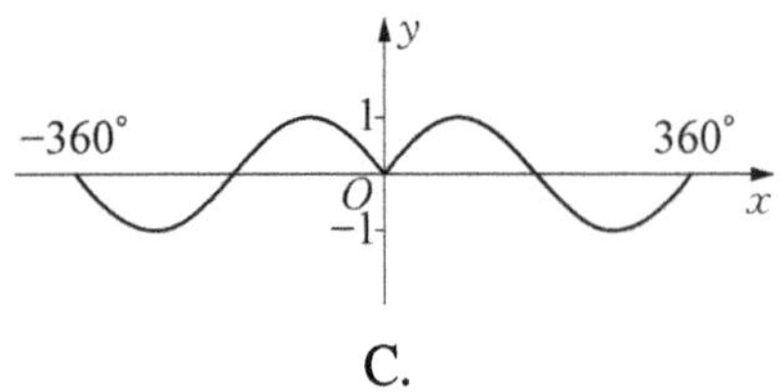

C.

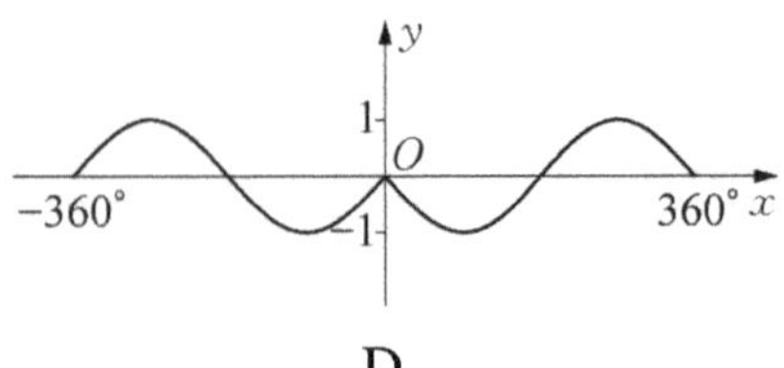

D.

3 How many points do the graphs of the functions $y = \sin x$ and $y = \tan x$ have in common in the interval $0° \leqslant x \leqslant 360°$? (   )

A. 1 B. 2 C. 3 D. 4

B. Fill in the blanks

4 Given the function $y = 1 - 2\sin x$, the set of values that y can take is ________.

5 One of the symmetrical points of the graph of the function $y = \tan x$ is ________.

6 Given the function $f(x) = \cos(x \times 60°)$ and that x is an integer, then the set of values that $f(x)$ can take is ________.

7 Given the function $y = \tan x$ $(-45° \leqslant x \leqslant 45°)$, the set of values that y can take is ________.

C. Questions that require solutions

8 Draw the graph of each of the following functions.

(a) $y = 1 + \cos x$ $(0° \leqslant x \leqslant 360°)$ (b) $y = 1 - \sin x$ $(0° \leqslant x \leqslant 360°)$

9 In the same coordinate plane, sketch the graphs of the function $y = \cos x$ $(-90° \leqslant x \leqslant 270°)$ and $y = 1 + \sin x$ $(0° \leqslant x \leqslant 360°)$. Describe how the graph of $y = \cos x$ $(-90° \leqslant x \leqslant 270°)$ can be translated to obtain the graph of $y = 1 + \sin x$ $(0° \leqslant x \leqslant 360°)$.

10 Sketch the graph of $y = \frac{1}{2} + \sin x$ $(-180° \leqslant x \leqslant 180°)$ and find the set of values of x when:

(a) $y > 0$

(b) $y \leqslant 1$.

4.10 Trigonometric functions (2)

Learning objective

Recognise, sketch and interpret graphs of the trigonometric functions $y = \sin x$, $y = \cos x$, and $y = \tan x$ for angles of any size.

A. Multiple choice questions

1. Which of these is the graph of the function $y = \sin(-x)$ $(-180° \leqslant x \leqslant 180°)$? (   )

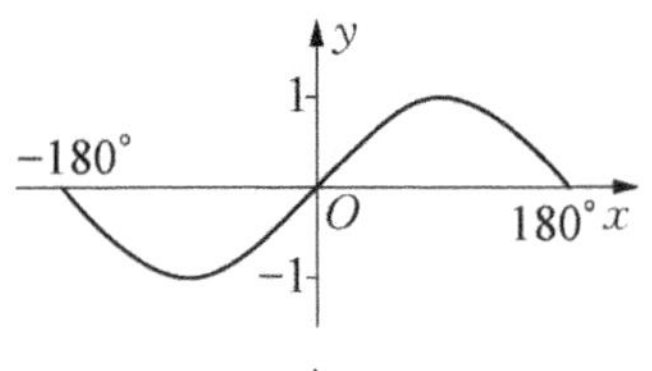

A.

B.

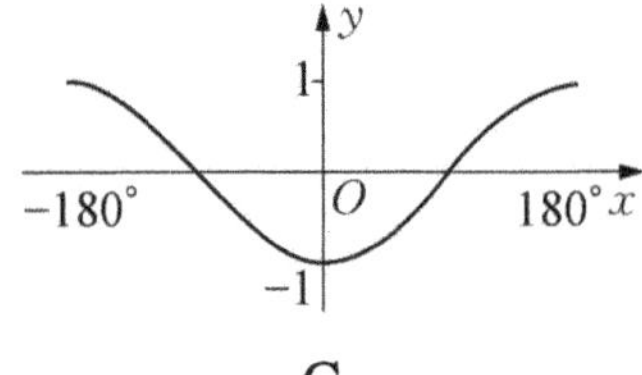

C.

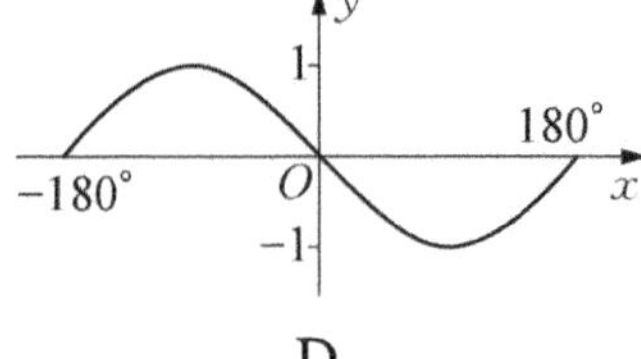

D.

2. Which of these is the graph of the function $y = \cos(-x)$ $(-360° \leqslant x \leqslant 360°)$? (   )

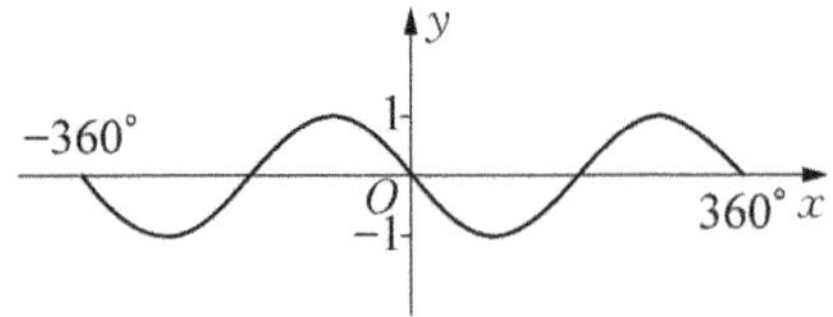

A.

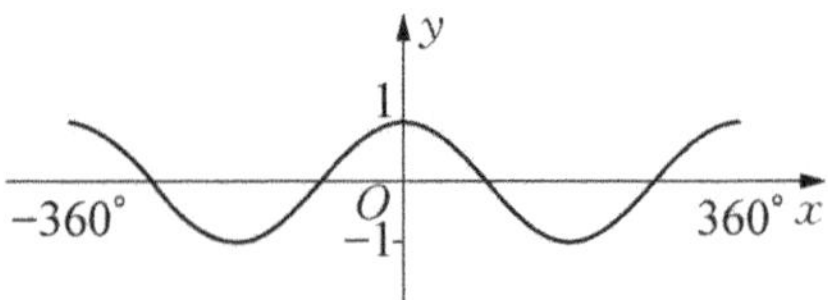

B.

C.

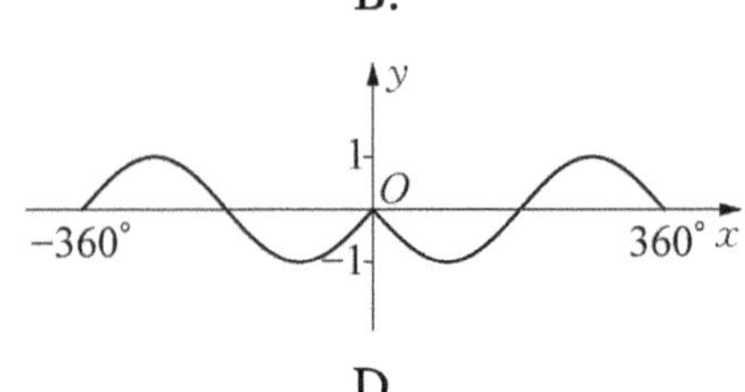

D.

3. Given the function $y = \sin x + \tan x$, which of the following statements is incorrect? (   )

A. The function has no maximum value.
B. The function has no minimum value.
C. The function has no maximum nor minimum value.
D. The function has a maximum value.

B. Fill in the blanks

4 The maximum value of $y = \sin x$ is ________.

5 When $y = 3 + \cos x$ reaches its minimum value, $x =$ ________.

6 When $y = 2 - \sin x$ reaches its maximum value, $x =$ ________.

C. Questions that require solutions

7 Given that the function $y = a + b\cos x$ has its maximum value $\frac{3}{2}$ and minimum value $-\frac{1}{2}$, find the values of the real numbers a and b.

8 Find the maximum and minimum values of the function $f(x) = \cos^2 x + \sin x - 1$.

9 Find the maximum and minimum values of the function $y = \tan^2 x - 3\tan x + 2$, when $-45° \leqslant x \leqslant 60°$.

10 Given that the equation $\cos x = 2a + 1$ does not hold true, then find the set of the values that the real number a can take.

Unit test 4

A. Multiple choice questions

1 Given that A is an acute angle and $\sin A = \frac{1}{2}$, then $\cos A =$ ( ).

A. 1 B. $\frac{1}{2}$ C. $\frac{\sqrt{3}}{2}$ D. $\frac{\sqrt{2}}{2}$

2 If an angle α in a triangle satisfies the equation $\sqrt{2}\sin(\alpha - 35°) = 1$, then α is a/an ( ).

A. obtuse angle B. acute angle

C. right angle D. obtuse angle or acute angle

3 If α is an angle of a equilateral triangle, then the value of $\tan\alpha$ is ( ).

A. $\sqrt{3}$ B. 1 C. $\frac{\sqrt{3}}{3}$ D. $\frac{\sqrt{2}}{2}$

4 In a triangle ABC, given that $\sqrt{\sin A - \frac{1}{2}} + (\sqrt{3}\tan B - 3)^2 = 0$, then the triangle ABC is a/an ( ).

A. obtuse-angled triangle B. acute-angled triangle

C. right-angled triangle D. not sure

5 The diagram shows a simple method to measure the distance between two points A and B on the opposite sides of the river. Point C is such that PC is perpendicular to AB. If $BC = 15\,m$ and $\angle ACB = 48°$, then AB is ( ).

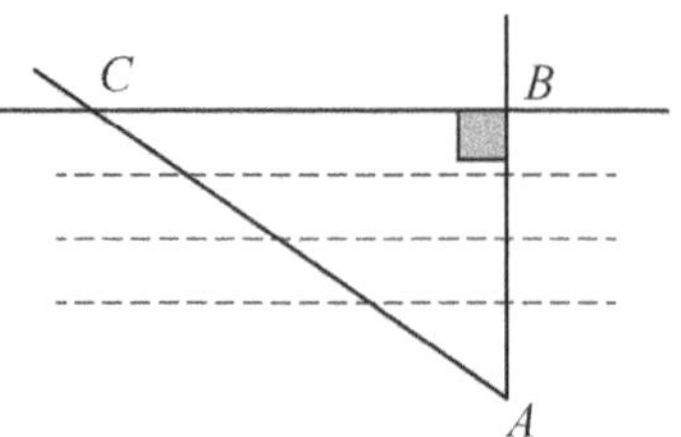

A. $15\sin 48°$ B. $15\cos 48°$

C. $\frac{15}{\tan 48°}$ D. $15\tan 48°$

6 The diagram shows a rectangle $ABCD$ with a triangle ADE inside it. $AD = 10\,\text{cm}$, E is a point on side BC such that ED is the angle bisector of angle $\angle AEC$ and $\sin\angle AEB = \frac{3}{5}$. Then $CE =$ (　　).

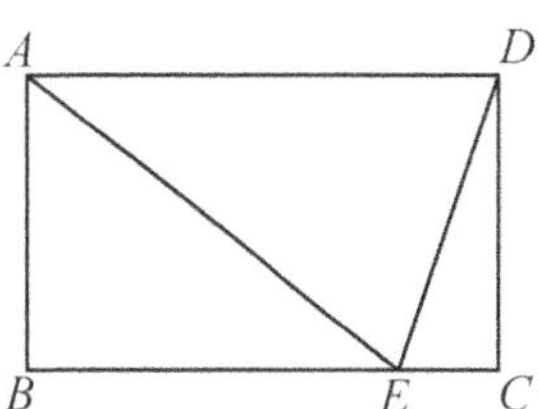

A. 2　　B. 4

C. 6　　D. 8

7 $\sin 2022°$ is equivalent to (　　).

A. $\sin 42°$　　B. $-\sin 42°$　　C. $\cos 42°$　　D. $-\cos 42°$

8 In a right-angled triangle ABC, angle $C = 90°$, $\sin A = \frac{1}{3}$ and $BC = 2\,\text{cm}$. Then the length of AB is (　　).

A. $\frac{2}{3}$ cm　　B. 4 cm　　C. $4\sqrt{2}$ cm　　D. 6 cm

9 If α is an acute angle, the terminal side of $k \times 360° - \alpha$ ($k \in \mathbf{Z}$) lies in the (　　) quadrant.

A. first　　B. second　　C. third　　D. fourth

10 One of the lines of symmetry of the graph of the function $f(x) = \sin\left(2x + \frac{\pi}{3}\right)$ can be (　　).

A. $x = \frac{\pi}{12}$　　B. $x = \frac{5\pi}{12}$

C. $x = \frac{\pi}{3}$　　D. $x = \frac{\pi}{6}$

B. Fill in the blanks

11 Calculate: $2\sin 30° + (\pi - 2022)^0 + |\sqrt{2} - 1| + \left(\frac{1}{2}\right)^{-1} - 2\sin 45° =$ ________.

12 In a right-angled triangle ABC, $\angle C = 90°$, and $\tan A = \frac{4}{3}$. Then, $\sin A + \cos A =$ ________.

13 Angles whose terminal sides coincide with that of angle $-600°$ are all in the ________ quadrant. The smallest positive of such angle is ________ and the largest negative of such angle is ________.

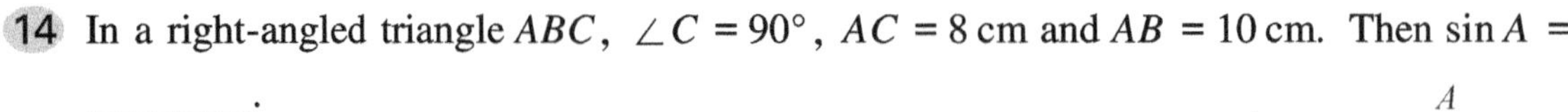

14 In a right-angled triangle ABC, $\angle C = 90°$, $AC = 8$ cm and $AB = 10$ cm. Then $\sin A =$ ________.

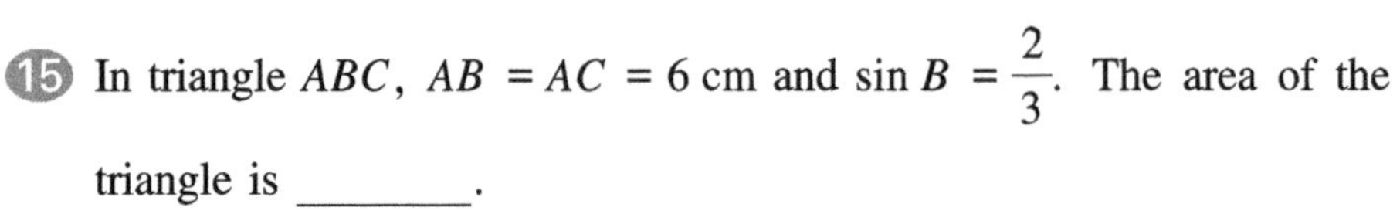

15 In triangle ABC, $AB = AC = 6$ cm and $\sin B = \frac{2}{3}$. The area of the triangle is ________.

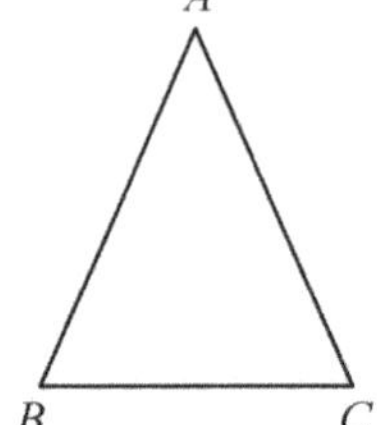

16 Given the function $y = \cos x$ $(30° \leqslant x \leqslant 135°)$, the set of values that y can take is ________.

17 Given that the maximum value of the function $y = a\sin x + 2$, is 3, then the value of a is ________.

18 Given the function $f(x) = \sin\left(x + \frac{\pi}{6}\right)$ for $-\frac{\pi}{3} \leqslant x \leqslant a$. When $a = \frac{\pi}{2}$, the set of values that y can take is ________. If the set of values that y can take is $-\frac{1}{2} \leqslant y \leqslant 1$, then the set of values that a can take is ________.

C. Questions that require solutions

19 The diagram shows a right-angled triangle ABC in which $\angle C = 90°$, $AC = 4$ cm and $AB = 5$ cm. Point D is on AC such that $AD = 3$ cm and DE is perpendicular to AB. Find the length of side AE.

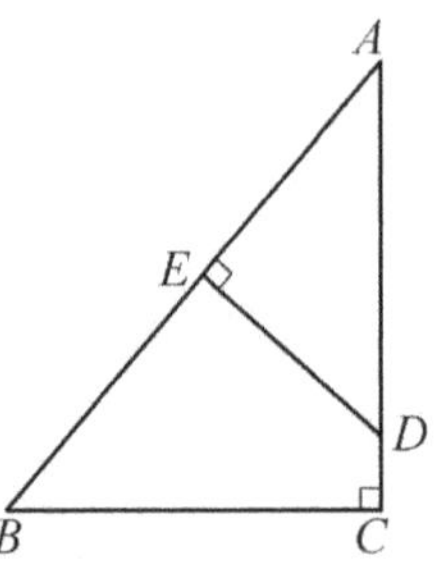

20 Look at the diagram. Given that $BM = 1.6\,\text{m}$, $BC = 13\,\text{m}$, $\angle ABC = 35°$ and $\angle ACE = 45°$, find the length of side AD.

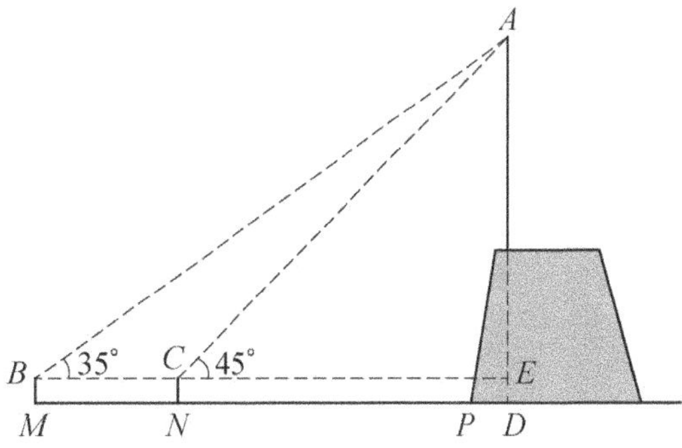

21 Given that $\sin\theta$ and $\cos\theta$ are the two roots of the equation $2x^2 - bx + \frac{1}{4} = 0$ for $\frac{\pi}{4} < \theta < \frac{3\pi}{4}$, answer the following questions.

(a) Find the value of b.

(b) Find the value of $\frac{2\sin\theta\cos\theta + 1}{\cos\theta - \sin\theta}$.

22 The diagram shows a trapezium $ABCD$ in which AD is parallel to BC, angle $A = 90°$, $BD = BC$ and CE is perpendicular to BD.

(a) Prove that $BE = AD$.

(b) Given that angle $DCE = 15°$ and $AB = 2\text{ cm}$, find the area of the trapezium $ABCD$.

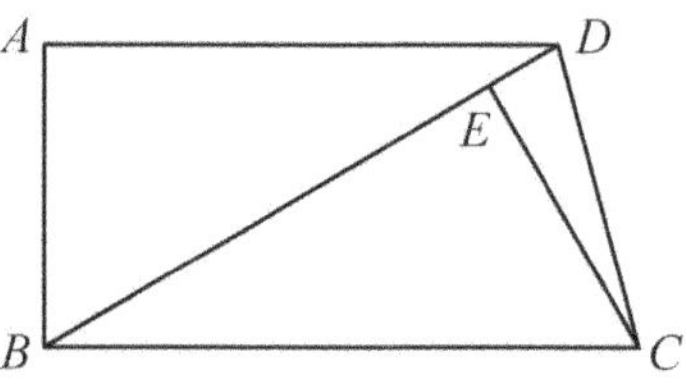

23 (a) Sketch the graph of the function $f(x) = \sin x + 3$ for $0° \leqslant x \leqslant 360°$.

(b) Describe how the graph of $f(x) = \sin x$ $(0° \leqslant x \leqslant 360°)$ can be translated to obtain the graph of $f(x) = \sin x - 2$.

(c) Hence, find the value of x for $\cos^2 x = \sin x - 2$, $0° \leqslant x \leqslant 360°$.

(d) Find the maximum and minimum values of the function $y = \cos^2 x + \sin x + 1$, when $30° \leqslant x \leqslant 120°$.

Chapter 5 Sequences

5.1 Revision of sequences

Learning objective

Generate terms of a sequence from either a term-to-term or a position-to-term rule.

A. Multiple choice questions

1 The general term of the sequence 1, 0, 1, 0, 1, ⋯, is (   ).

A. $a_n = \frac{1-(-1)^{n+1}}{2}$ B. $a_n = \frac{1+(-1)^{n+1}}{2}$

C. $a_n = \frac{(-1)^n - 1}{2}$ D. $a_n = \frac{-1-(-1)^n}{2}$

2 A sequence is given by 1, $\sqrt{3}$, $\sqrt{5}$, $\sqrt{7}$⋯, $\sqrt{2n-1}$, ⋯. Then $3\sqrt{5}$ is its (   ).

A. 22nd term B. 23rd term C. 24th term D. 28th term

3 In the sequence 11, 13, 15, ⋯, $2n+1$, there are (   ) terms.

A. n B. $n-3$ C. $n-4$ D. $n-5$

4 Which of the following is not the general term of the sequence 2, 4, 8, ⋯? (   )

A. $a_n = 2^n$. B. $a_n = n^2 - n + 2$

C. $a_n = -\frac{2}{3}n^3 + 5n^2 - \frac{25}{3}n + 6$ D. $a_n = 2n$

B. Fill in the blanks

5 Write down the first five terms of the sequence $a_n = 2n$: ________________.

6 Look at each of the following pattern sequences. Draw the next pattern in each sequence and write the general formula for the total number of shapes in each pattern.

________________; and the general formula is ________.

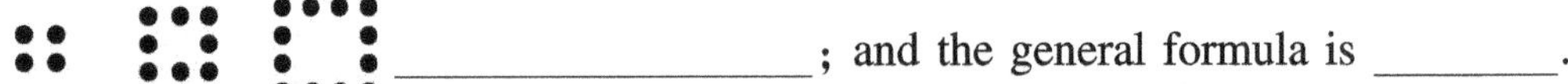
____________; and the general formula is ________.

7 The general formula of an arithmetic sequence $\{a_n\}$ is $a_n = 5 - 4n$. The common difference of the sequence $d =$ ________.

8 In a geometric sequence $\{c_n\}$, $c_9 = 1$ and $c_7 = 4$. Then its common ratio $q =$ ________.

9 Given that the first term of a geometric sequence $\{a_n\}$ is 3 and the common ratio is $\frac{1}{2}$, then the general formula of the sequence is $a_n =$ ________.

10 Given that $a_n = \frac{1}{n(n+1)}$ is the general formula of the sequence $\{a_n\}$, then $\frac{1}{30}$ is the ________ term of the sequence.

C. Questions that require solutions

11 Assuming that the general formula of a sequence $\{a_n\}$ is $a_n = -3n + 10$, find the sum of the first ten terms whose values are positive in the sequence.

12 It is given that the general formula of a sequence $\{a_n\}$ is $a_n = \frac{3n-2}{3n+1}$.

(a) Find the 10th term of the sequence.

(b) Is $\frac{98}{101}$ a term of this sequence?

(c) Prove that the terms of this sequence are all in the interval $0 < a_n < 1$.

(d) Are there any terms of the sequence in the interval $\frac{1}{3} < a_n < \frac{2}{3}$? If so, state how many terms there are in this interval. If there are no terms, give a reason.

5.2 Arithmetic sequences

Learning objective

Recognise and use simple arithmetic sequences.

A. Multiple choice questions

1 An arithmetic sequence satisfies the equation $a_1 + a_2 + a_3 + \cdots + a_{101} = 0$. Then (　　).

A. $a_1 + a_{101} > 0$　　B. $a_2 + a_{100} < 0$　　C. $a_3 + a_{99} = 0$　　D. $a_{51} = 51$

2 The sequences x, a_1, a_2, y and x, b_1, b_2, b_3, y are both arithmetic sequences and $x \neq y$. Then the value of $\dfrac{b_2 - b_1}{a_2 - a_1}$ is (　　).

A. $\dfrac{2}{3}$　　B. $\dfrac{3}{4}$　　C. $\dfrac{4}{3}$　　D. $\dfrac{3}{2}$

3 Given that a sequence, a_1, a_2, a_3, $\cdots$, a_n, $\cdots$, is an arithmetic sequence with a common difference of d, then the sequence, $a_1 + a_3$, $a_2 + a_4$, $a_3 + a_5$, $\cdots$, $a_n + a_{n+2}$, $\cdots$, is (　　).

A. an arithmetic sequence with a common difference of d

B. an arithmetic sequence with a common difference of $2d$

C. an arithmetic sequence with a common difference of $3d$

D. not an arithmetic sequence

B. Fill in the blanks

4 Given that the sequence $\{a_n\}$ is an arithmetic sequence, $a_3 = -6$ and $a_6 = 0$, then its general formula $a_n =$ ________.

5 Given a sequence $\{a_n\}$ with $a_1 = 3$ and $a_{n+1} = a_n - 2$, then $a_8 =$ ________.

6 In an arithmetic sequence $\{a_n\}$, if $a_4 + a_6 + a_8 + a_{10} + a_{12} = 120$, then $2a_9 - a_{10} =$ ________.

7 Given an arithmetic sequence 7, 2, −3, ⋯, then the common difference of the sequence is ________, the 20th term of this sequence is ________ and −98 is the ________ term of this sequence.

8 Between 10 and 100, there are ________ integers that are divisible by 7.

9 Given that the general formula of a sequence $\{a_n\}$ is $a_n = 3n - 2$, and a new sequence is formed by a_1, a_4, a_7, a_{10}, ⋯, then the general formula of the new sequence is b_n = ________.

C. Questions that require solutions

10 It is given that a sequence $\{a_n\}$ is an arithmetic sequence, $a_6 = 4$ and $a_{14} = 64$. Let the arithmetic mean of a_6 and a_{14} be x and the arithmetic mean of a_6 and x be y. Find the value of $x + y$.

11 It is given that a sequence $\{a_n\}$ is an arithmetic sequence with common difference d. Let $b_n = a_{2n-1} + a_{2n}$. Prove that $\{b_n\}$ is an arithmetic sequence and find the common difference of $\{b_n\}$.

12 In an arithmetic progression $\{a_n\}$, $a_6 + a_7 + a_8 = 17$ and $a_4 + a_5 + a_6 + \cdots + a_{14} = 77$. If $a_k = 13$, find the value of k.

13 A number is to be filled in each grid of the square so that the five numbers in each row, from left to right, form an arithmetic sequence with a common difference of x, and the five numbers in each column, from top to bottom, form an arithmetic sequence with common difference y. Three numbers are shown in the square. What number should be filled in the upper left corner of the grid?

			0	
	9			
				18

5.3 Geometric sequences

Learning objective

Recognise and use geometric sequences.

A. Multiple choice questions

1 The common ratio of a geometric sequence (or geometric progression) a_n is $q(q \neq 1)$. Let $b_1 = a_1 + a_2 + a_3$, $b_2 = a_4 + a_5 + a_6$, $\cdots$ and $b_n = a_{3n-2} + a_{3n-1} + a_{3n}$, $\cdots$. Then the sequence $\{b_n\}$ is ( ).

A. an arithmetic sequence

B. a geometric sequence with common ratio $q(q \neq 1)$

C. a geometric sequence with common ratio q^3

D. is neither an arithmetic sequence nor a geometric sequence

2 A geometric sequence that is decreasing and has first term a_1 and common difference q satisfies condition ( ).

A. $a_1 < 0$, $q > 1$ or $a_1 > 0$, $0 < q < 1$ B. $a_1 < 0$, $0 < q < 1$ or $a_1 > 0$, $1 < q$

C. $a_1 < 0$, $q > 0$ or $a_1 > 0$, $q < 0$ D. $a_1 < 0$, $q > 1$ or $a_1 > 0$, $q < 1$

3 Given that an arithmetic sequence $\{a_n\}$ has a common difference of $d \neq 0$, and a_2, a_3, a_6, in order, constitute a geometric sequence, then the common ratio of this geometric sequence is ( ).

A. $\frac{1}{2}$ B. $\frac{1}{3}$ C. 2 D. 3

B. Fill in the blanks

4 In a sequence $\{a_n\}$, $a_1 = 3$ and $a_{n+1} = -\frac{1}{3}a_n$. Then $a_n =$ ________.

5 In a geometric sequence $\{b_n\}$, $b_7 = 1$ and the common ratio is $q = \frac{1}{7}$. Then its general formula is $b_n =$ ________.

6 Given that in a geometric sequence $\{a_n\}$, $a_3 + a_7 = 56$, and $a_4 + a_8 = 28$, then its common ratio is $q =$ ________.

Geometric mean is the nth root of the product of n positive numbers. Given positive numbers a and b, their geometric mean is $\sqrt{ab}$.

7 The arithmetic mean of $2 - \sqrt{3}$ and $2 + \sqrt{3}$ is ________, and the geometric mean is ________.

8 In a geometric sequence $\{a_n\}$, if $a_3 a_{13} = 36$ and $a_7 = 9$, then $a_9 =$ ________.

9 If the common ratio of a geometric sequence $\{a_n\}$ is a real number, $a_2 = 4$ and $a_6 = 64$, then $a_3 + a_4 + a_5 =$ ________.

C. Questions that require solutions

10 Given that a sequence $\{a_n\}$ is a geometric sequence and $a_n = 2 \times 3^n + a$, find the value of the real number a.

11 It is given that a positive sequence $\{a_n\}$ is a geometric sequence with common ratio q. Prove that: (a) the sequence $\{a_{2k+1}\}$ (k is a natural number) is a geometric sequence (b) the sequence $\{\sqrt{a_n}\}$ is a geometric sequence.

12 Given that the sequence $\{a_n\}$ is an arithmetic sequence, its common difference is $d \neq 0$ and a_1, a_3, a_9 in order constitute a geometric sequence, find the value of $\frac{a_1 + a_3 + a_9}{a_2 + a_4 + a_{10}}$.

13 Given that each term of a geometric sequence is a real number, the product of the first four terms is 81, and the sum of the second term and the third term is 10, find the value of common ratio of this geometric sequence.

5.4 Fibonacci sequences

Learning objective

Recognise and use Fibonacci-type sequences.

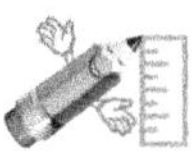

A. Multiple choice questions

1 The animals that Fibonacci was studying when he discovered the number sequence named after him were ( ).

A. rabbits B. dogs C. cows D. sheep

2 Which of the following lists of ordered numbers can form part of a Fibonacci-type sequence? ( )

A. 1, 2, 3, 4, 5, 6
B. 1, 1, 2, 3, 5, 8
C. 1, 2, 8, 16, 32, 64
D. 0, 0, 1, 1, 2, 3

3 If 2, 3, 5, x, 13, 21, ⋯ is part of a Fibonacci-type sequence, then x is ( ).

A. 6
B. 7
C. 8
D. any integer

4 If 8, 13, 21, x, 55, 89, y, ⋯ is part of the Fibonacci sequence, then the sum of x and y is ( ).

A. 112
B. 178
C. 186
D. none of the above

B. Fill in the blanks

5 A sequence $\{a_n\}$ satisfies $a_1 = 3$, $a_2 = 5$ and $a_n = a_{n-1} + a_{n-2} (n \geqslant 3, n \in \mathbf{N})$, where **N** is the set of natural numbers. Then $a_6 =$ ________.

6 There are six steps in a staircase. You can ascend the staircase by going up by either one step or two steps at a time. There are ________ ways of walking from the ground level to the top of the stairs.

7 For a sequence $\{a_n\}$, $a_1 = 1$ and $a_n = 1 + \frac{1}{a_{n-1}}(n \geqslant 2, n \in \mathbf{N})$. The first five terms of this sequence are ________, ________, ________, ________, ________.

8 For a given sequence, $a_1 = 1$, $a_2 = 1$ and $a_{n+2} = a_{n+1} + a_n(n \in \mathbf{N})$. Then $\frac{a_1^2 + a_2^2 + \cdots + a_{2019}^2}{a_{2019}}$ is the ________ term of this sequence.

C. Questions that require solutions

9 In the Fibonacci sequence 1, 1, 2, 3, 5, 8, $\cdots$, if $a_{2016} = a$, find the value of $a_1 + a_2 + \cdots + a_{2014}$.

10 In the Fibonacci sequence 1, 1, 2, 3, 5, 8, $\cdots$, let the 2018th term be m. What is the difference between the 2019th term and the sum of the first 2018 terms? Express your answer in terms of m.

11 Let α and β be the two roots of the quadratic equation $x^2 - x - 1 = 0$, and let the sequence $\{a_n\}$ satisfy $a_n = \frac{\alpha^n - \beta^n}{\alpha - \beta}(n = 1, 2, 3, \cdots)$. Prove that, for any positive integer n, $a_{n+2} = a_{n+1} + a_n$.

5.5 More about sequences

Learning objective

Recognise and use sequences of triangular, square and cube numbers, quadratic and other sequences.

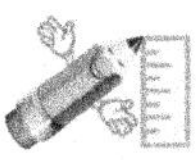

A. Multiple choice questions

1. Which of the following list of ordered numbers forms part of the sequence of square numbers? (　　)

A. 1, 2, 3, 4, 5　　B. 1, 9, 16, 25, 36

C. 0, 1, 4, 16, 256　　D. 9, 16, 25, 36, 49

2. The pattern is the start of the sequence of triangular numbers. The number of dots in the next triangle is (　　).

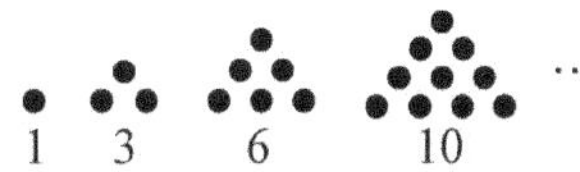

A. 11　　B. 13　　C. 15　　D. 16

3. The 6th term in the sequence of cube numbers, 1, 8, 27, 64, 125, ⋯, is (　　).

A. 189　　B. 250

C. 216　　D. none of the above

4. Which of the following numbers is both in the sequence of square numbers and the sequence of cube numbers? (　　)

A. 100　　B. 1000　　C. 10 000　　D. 1 000 000

B. Fill in the blanks

5. The 8th term of the sequence of square numbers is ________, and the 8th term of the sequence of cube numbers is ________.

6. The 10th term in the sequence of triangular numbers is ________.

7 5050 is a term in the sequence of ________ numbers. (Choose from: square cube triangular.)

8 The sum of the 5th term of the sequence of square numbers and the 5th term of the sequence of cube numbers is ________.

9 When n increases, the difference between the nth term of the sequence of square numbers and the nth term of the sequence of cube numbers ________. (Choose from: increases decreases does not change.)

C. Questions that require solutions

10 Given the triangular number sequence, 1, 3, 6, 10, 15, 21, ⋯, answer the following.

(a) Write the 10th term.

(b) Write the formula for the general term, or nth term, a_n.

(c) Let all the triangular numbers that are divisible by 5 form a number sequence $\{b_n\}$. Find the formula for its general term, b_n, and hence the value of b_{2019}. What is the position of this value in the original sequence of triangular numbers $\{a_n\}$?

11 Given that a sequence $\{a_n\}$ satisfies $a_1 = \frac{1}{4}$ and $\frac{1}{a_{n+1}} = \frac{1}{a_n} + 3$, find the exact value of a_5.

12 Prove that, when n is an even number, the sum of the three nth terms in the sequence of square numbers, the sequence of cube numbers and the sequence of triangular numbers is divisible by $n + 1$.

5.6 Calculate the *n*th term of linear and quadratic sequences

Learning objective

Deduce expressions to calculate the *n*th term of linear and quadratic sequences.

A. Multiple choice questions

1 Which of the following statements is incorrect? (   )

A. A linear sequence is also called an arithmetic sequence.

B. A linear sequence is either increasing or decreasing from each term to its next term.

C. A linear sequence cannot be also a quadratic sequence.

D. A square sequence is a quadratic sequence.

2 Of the following ordered numbers, (   ) forms part of a linear sequence.

A. 1, 20, 30, 40, 50   B. 2, 4, 8, 16, 32

C. 0, 1, 0, 1, 0   D. 100, 50, 0, −50, −100

3 Given that the *n*th term of a linear sequence is $a_n = 3n - 5$, then its 100th term is (   ).

A. 100   B. 300   C. 295   D. 305

4 Given that the *n*th term of a quadratic sequence is $a_n = n^2 + 2n - 15$, then the term whose value is 0 is its (   ) term.

A. first   B. second   C. third   D. fourth

B. Fill in the blanks

5 The first five terms of a linear sequence are −5, 0, 5, 10, 15. The general formula for the *n*th term of the sequence is a_n = ________.

6 The general formula of a linear sequence $\{a_n\}$ is $a_n = 3n - 13$. There are ________ terms in the sequence whose values are positive.

7 For a given sequence $\{a_n\}$, $a_1 = 3$ and $a_{n+1} - a_n = 6$. Then a_n = ________.

8 The general formula of a quadratic sequence $\{a_n\}$ is $a_n = 5n^2 - 8$. Then $a_6 =$ ________.

9 The first five terms of a quadratic sequence are 0, 3, 8, 15, 24. Then the general formula for the nth term of the sequence is $a_n =$ ________.

C. Questions that require solutions

10 The first five terms of a linear sequence are −18, −15, −12, −9, −5. Find an expression for the nth term of the sequence.

11 The general term of a linear sequence $\{a_n\}$ is $a_n = 2n - 3$. Some of the terms of $\{a_n\}$ are used to form a new linear sequence $\{b_n\}$ without changing the order. Suppose $b_1 = a_2$, and a_6 is a term of the new sequence $\{b_n\}$. Find the general formula of the sequence $\{b_n\}$.

12 The first six terms of a sequence are 2, 7, 14, 23, 34, 47.

(a) Explain why these numbers are not part of a linear sequence.

(b) Explain why these numbers could be part of a quadratic sequence.

(c) Suppose the numbers given form part of a quadratic sequence. Write the general formula for the nth term of the sequence.

Unit test 5

A. Multiple choice questions

1 The first term of an arithmetic sequence is −12 and the 12th term is 10. The common difference of the sequence d = ( ).

A. 2 B. $\frac{3}{2}$ C. $\frac{1}{2}$ D. −2

2 The sequence $\{a_n\}$ is an increasing arithmetic sequence and a_2 and a_4 are the roots of equation $x^2 + 6x + 5 = 0$. Then a_5 = ( ).

A. 7 B. 3 C. 1 D. −1

3 The first three terms of an arithmetic sequence $\{a_n\}$, are $a - 1$, $a + 1$ and $a + 3$. The formula for the general term of the sequence is ( ).

A. $a_n = 2n - 3$ B. $a_n = 2n - 1$ C. $a_n = a + 2n - 3$ D. $a_n = a + 2n - 1$

4 In a geometric sequence $\{a_n\}$, $a_1 + a_2 + a_3 = 1$ and $a_2 + a_3 + a_4 = 2$. The common ratio r of the sequence is ( ).

A. −3 B. 2 C. −2 D. 4

5 In an arithmetic sequence $\{a_n\}$ with a non-zero common difference, $a_2 = 4$, and a_1, a_2 and a_6 are three consecutive terms of a geometric sequence. Then the common difference of the arithmetic sequence d = ( ).

A. 3 B. 4 C. 5 D. 6

6 In a geometric sequence $\{a_n\}$, $a_1 + a_2 = 1$ and $a_3 + a_4 = 2$. Then $a_{15} + a_{16}$ = ( ).

A. 32 B. 64 C. 128 D. 256

7 Different numbers of wooden sticks are used to form a "goldfish" pattern sequence, as shown in the diagram. How many sticks does the nth goldfish need?

   …

① ② ③

A. $4n+4$ B. $8n$ C. $6n+2$ D. $10n-2$

8 Consider the Fibonacci sequence: 1, 1, 2, 3, 5, 8, 13, ⋯. Each number in this sequence is used as the side length of a square, and then the squares are used to make a sequence of rectangles, as shown in the diagram. The perimeter of the rectangle with pattern number ⑦ is ( ).

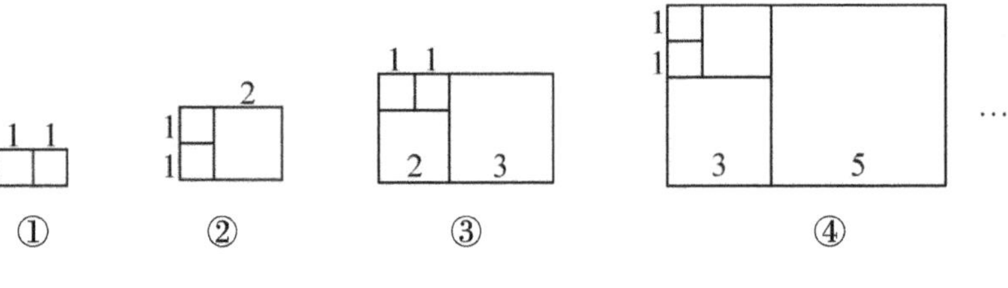

A. 68 B. 42 C. 110 D. 178

9 Given that $\{a_n\}$ is an arithmetic sequence, and $\{b_n\}$ is a geometric sequence in which all the terms are positive and the common ratio $r>1$. If $a_4=b_4$, then ( ).

A. $a_2+a_6>b_3+b_5$ B. $a_2+a_6=b_3+b_5$

C. $a_2+a_6<b_3+b_5$ D. not sure

10 The Fibonacci sequence can be defined by recursion as follows: using a_n to represent the general term, the sequence $\{a_n\}$ satisfies: $a_1=1$, $a_2=1$ and $a_n=a_{n-1}+a_{n-2}(n\geqslant 3)$. Which of the following statements is correct? ( )

① $a_3=2$ ② $3a_n=a_{n-2}+a_{n+2}(n\geqslant 3)$ ③ $\sum_{i=1}^{2021} a_i=a_{2023}$ ④ $\sum_{i=1}^{2021} a_i^2=a_{2021}a_{2022}$

A. ①②③ B. ①②④

C. ②③④ D. ①③④

B. Fill in the blanks

11 Consider the sequence 3, 9, 15, 21, 27, ⋯. Then the 99th term of the sequence is ________, and 489 is the ________ term of the sequence.

12 Given that $k+5$, -1 and $2k-1$ are consecutive terms of an arithmetic sequence, then $2k-1$ is ________.

13 The general term of the geometric sequence: 2, $2\sqrt{2}$, 4, $4\sqrt{2}$, ⋯, is ________.

14 Given that the first four terms of an arithmetic sequence are 17, 24, 31 and 38, the first term of the sequence to exceed 2022 is ________.

15 The sum of the 8th term of the sequence of square numbers and the 6th term of the sequence of cube numbers is ________.

16 The 22nd term of the sequence of triangular numbers is ________. The formula for the general term of the sequence of triangular numbers is ________.

17 For a sequence $\{a_n\}$, $a_1 = 1$ and $a_n = a_{n-1} + \frac{1}{a_{n-1}}$ $(n > 2)$. The first four terms are ________, ________, ________, and ________.

18 All positive odd numbers can be arranged into a triangular array, as shown on the right. By observing the pattern, the 10th number on the 19th row is ________.

1
3 5
7 9 11
13 15 17 19
21 23 25 27 29
……

19 Given that the nth term of a quadratic sequence is $a_n = 2n^2 - 5n + 8$, then 11 is the ________ term.

20 The general formula of a linear sequence $\{a_n\}$ is $a_n = 14 - 6n$. There are ________ terms in the sequence whose values are negative.

21 A sequence $\{a_n\}$ satisfies $a_1 = a_2 = 1$ and $a_n = a_{n-1} + a_{n-2} (n \geqslant 3)$. Then $a_7 =$ ________.

C. Questions that require solutions

22 Consider the four numbers −6, −1, 4 and 9 in order.

(a) Show that the four numbers can form the first four terms of an arithmetic sequence.

(b) Find a formula for the general term of the arithmetic sequence.

(c) Find the 100th term of the sequence.

(d) Find the largest term in the sequence that is less than 500.

23 In a sequence $\{a_n\}$, $a_1 = 1$ and $a_{n+1} = -\frac{1}{a_n + 2}$,

(a) prove that the sequence $\left\{\frac{1}{a_n + 1}\right\}$ is an arithmetic sequence

(b) find a formula for the general term of $\{a_n\}$.

24 In Figure 1, the numbers of dots forming the triangles with pattern numbers (1), (2) and (3) are 3, 6 and 10, respectively. The pattern can be continued to form the sequence of triangular numbers. Similarly, in Figure 2, the numbers of dots forming the squares with pattern numbers (1), (2) and (3) are 4, 9 and 16, respectively. The pattern can also continue to form the sequence of square numbers.

(a) Write the 4th triangular number and the 4th square number.

(b) Write a formula for the nth triangular number and the nth square number.

(c) Prove that 36 is both a triangular number and a square number.

(d) Write two more numbers that are both triangular and square numbers.

Figure 1 **Figure 2**

Chapter 6 Two- and three-dimensional shapes

6.1 Enlargement of fractional and negative scale factors

Learning objective

Identify and describe similar shapes using scale factors and ratio notation; describe and calculate with scale diagrams and maps.

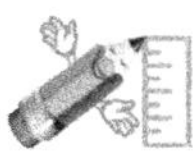

A. Multiple choice questions

1 The diagram shows an enlargement with centre O. Which of the following statements is incorrect? (   )

A. The scale factor of the enlargement from Figure I to Figure II is positive.

B. The Figure II scale factor of the enlargement from Figure I to Figure II is negative.

C. The scale factor of the enlargement from Figure I to Figure II is large than 1.

D. The scale factor of the enlargement from Figure I to Figure II is not 1.

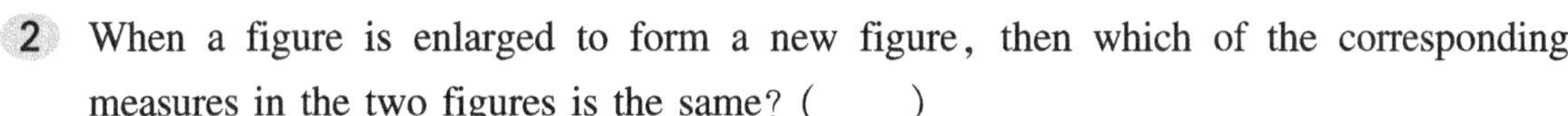

2 When a figure is enlarged to form a new figure, then which of the corresponding measures in the two figures is the same? (   )

A. the lengths of the sides　　B. the angles

C. the areas　　D. the perimeters

3 Given three line segments a, b and c, suppose that $\frac{1}{2}$ of the length of a is $\frac{1}{4}$ of the length of b and $\frac{1}{6}$ of the length of c. Then the ratio of the sum of the lengths of the three segments to the length of b is (   ).

A. 1 : 6　　B. 6 : 1

C. 1 : 3　　D. 3 : 1

4 Point P divides the line segment AB into two line segments with lengths in golden ratio, and $AP = \sqrt{5} - 1$. Then the length of AB is ( ).

A. 2 B. $\sqrt{5}+1$

C. 2 or $\sqrt{5}+1$ D. none of the above

5 In a school, there are 64 students in Year 7, and they are divided into three groups A, B and C. The ratio of the numbers of students in these groups is $4 : 5 : 7$. If a new student transferred from another school is now assigned to group B, then the new ratio of the number of students in group B to the number of students in group C is ( ).

A. $3 : 4$ B. $4 : 5$

C. $5 : 6$ D. $6 : 7$

B. Fill in the blanks

6 The actual distance between town A and town B is 24 kilometres. On a map with scale $1 : 800\,000$, the distance between the two towns is ________ cm.

7 If a triangle is enlarged with a scale factor of 10, then its area is enlarged to ________ times its original size. If its area is enlarged to 10 times its original size, then the lengths of its sides are enlarged to ________ times its original length.

8 In the following diagram, $AB : DB = AC : EC$, $AD = 15$ cm, and $AB = 40$ cm, $AC = 28$ cm. Then $AE =$ ________ cm.

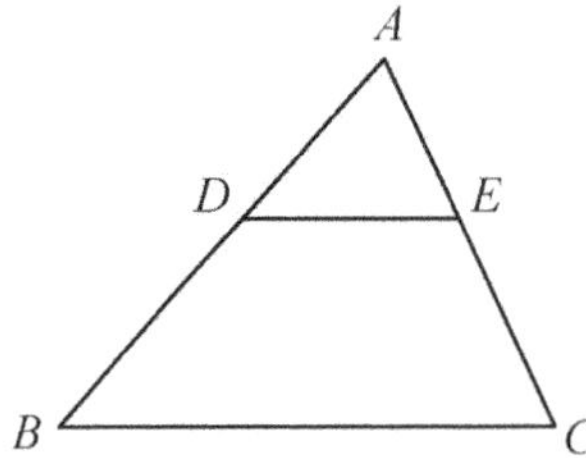

9 As shown in the diagram, triangle ABC inscribes a rhombus $AMPN$, where M, P, and N are on the sides of AB, BC and AC, respectively. If $\frac{AM}{MB} = \frac{1}{2}$, then $\frac{BP}{BC} =$ ________.

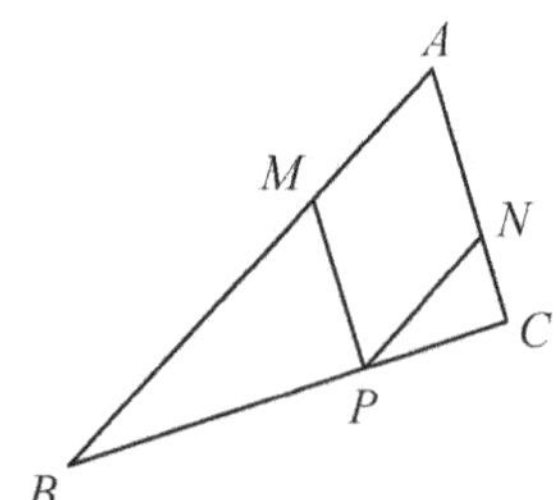

10 A "golden rectangle" is a rectangle in which the lengths of the adjacent sides are in the golden ratio. Visually, it is commonly regarded as the most beautiful rectangle. If the length of the longer side of a "golden rectangle" is 20 cm, then the length of its adjacent side is ________ cm.

11 The diagram shows a trapezium $ABCD$ with AD parallel to BC. AC and BD intersect at point O. If the area of $\triangle AOD =$ 4 square units and the area of $\triangle AOB = 6$ square units, then the area of $\triangle BOC =$ ________.

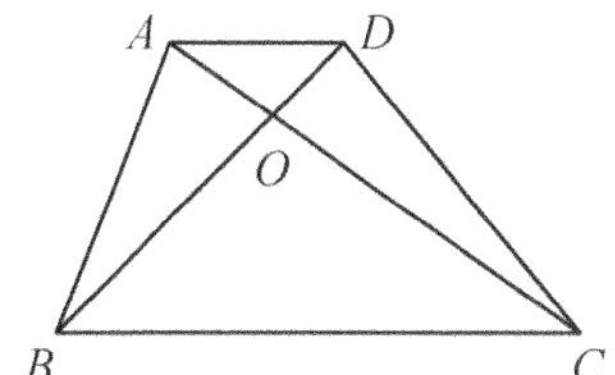

C. Questions that require solutions

12 On the coordinate plane shown below, using the origin as the centre of enlargement, draw the following figures:

(a) triangle $A'B'C'$ that is an enlargement of triangle ABC with a scale factor of $\frac{3}{2}$;

(b) triangle $A''B''C''$ that is an enlargement of triangle ABC with a scale factor of -2.

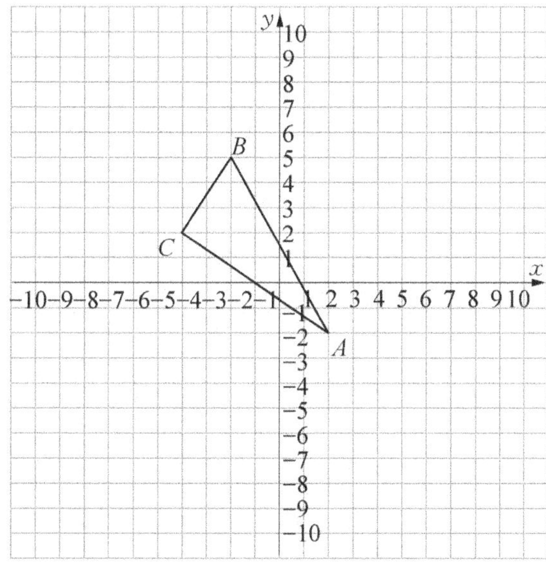

13 Suppose that a, b and c are the three sides of $\triangle ABC$, $a + b + c = 60$ cm and the ratio $a : b : c = 3 : 4 : 5$. Find the area of $\triangle ABC$ and the height perpendicular to the longest side.

14 On a map, the distance from town A to a school measures 3 cm, the distance from town B to the same school is 5 cm, and the actual distance from town B to the school is 10 kilometres. Find the actual distance from town A to the school.

15 In $\triangle ABC$, D is a point on BC. If $AB = 15$ cm, $AC = 10$ cm, $BD : DC = AB : AC$ and $BD - DC = 2$ cm, find the length of BC.

6.2 Combined transformation of geometric shapes

Learning objective

Describe the changes and invariance achieved by transformations and combinations of transformations.

A. Multiple choice questions

1 Look at the four designs below. Of the following statements, the correct one is ( ).

A.

B.

C.

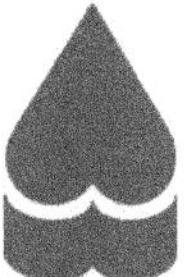

D.

A. Each figure can be viewed as a result of translating part of it.

B. Each figure can be viewed as a result of reflecting part of it.

C. Each figure can be viewed as a result of rotating part of it.

D. None of the above is correct.

2 Out of the following four figures, there is only one in which $\triangle A'B'C'$ can be obtained by reflecting $\triangle ABC$. This figure is ( ).

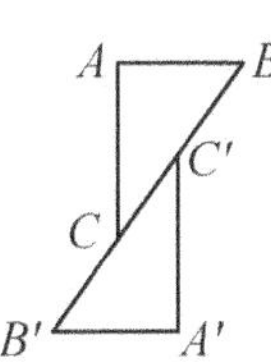

A.

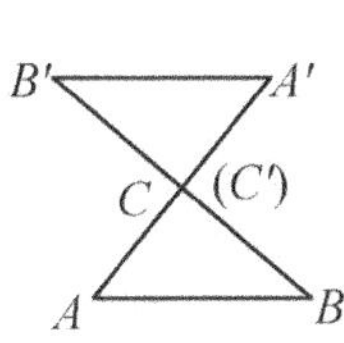

B.

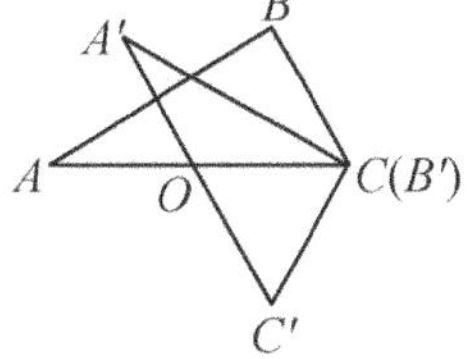

C.

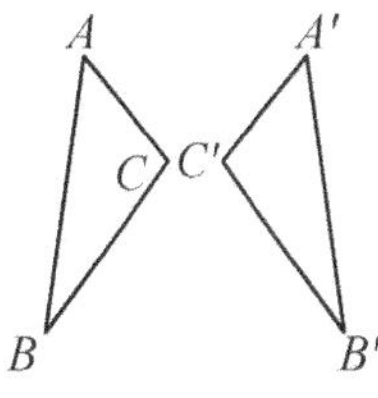

D.

3 A square $ABCD$ with side length 7 is translated in the direction of BC for 4 units to obtain the square $EFGH$. Then the length of AH is ( ).

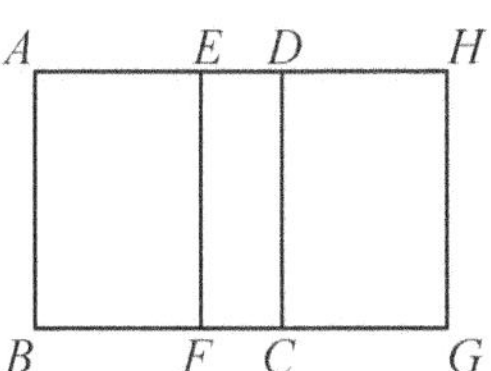

A. 9 B. 10

C. 11 D. 12

4 The diagram shows $\triangle ABC$, in which $\angle C = 90°$ and point D is on side AC. $\triangle BCD$ is folded along line BD such that point C falls at point E on the hypotenuse AB, and $DC = 5$ cm. Then the distance from D to the hypotenuse AB is ( ).

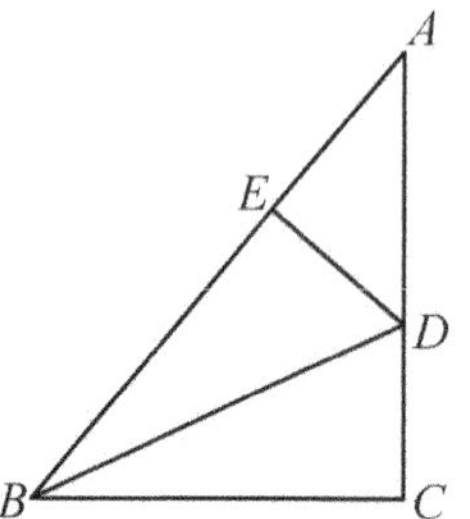

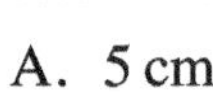

A. 5 cm B. 4 cm

C. 3 cm D. not sure

5 Which of the following statements is incorrect? ( )

A. After a rectangle is translated, it remains a rectangle, and its area and perimeter do not change.

B. If a rectangle is translated, it remains a rectangle, but its area and perimeter change.

C. If a rectangle is first translated and then rotated, the resulting figure remains a rectangle, and its area and perimeter do not change.

D. If a rectangle is first rotated and then reflected, the resulting figure remains a rectangle, and its area and perimeter do not change.

B. Fill in the blanks

6 A person stands in front of a mirror, and sees the time on the face of the clock on the opposite wall, as shown in the diagram. The actual time displayed on the clock is ________ : ________.

7 As shown in the diagram, $\triangle ABC$, $\triangle BDE$, $\triangle BCE$ and $\triangle CEF$ are all equilateral triangles with side length 2 cm. Translating $\triangle ABC$ to $\triangle BDE$, the distance of the translation is ________. Translating $\triangle ABC$ to $\triangle CEF$, the point corresponding to point A is ________, the line segment corresponding to AC is ________, and the angle corresponding to $\angle A$ is ________.

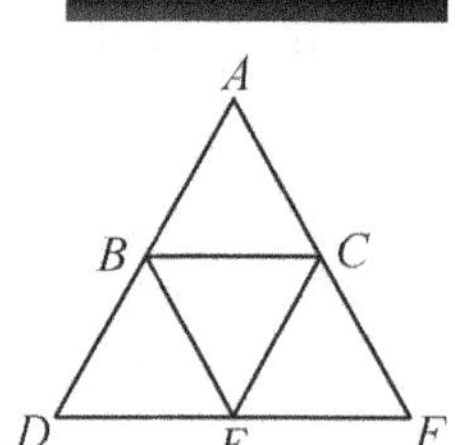

8 Look at the diagram. If $\triangle ABC$ with perimeter 8 cm is translated 1 cm in the direction of BC and the resulting figure is $\triangle DEF$, then the perimeter of the quadrilateral $ABFD$ is ________ cm.

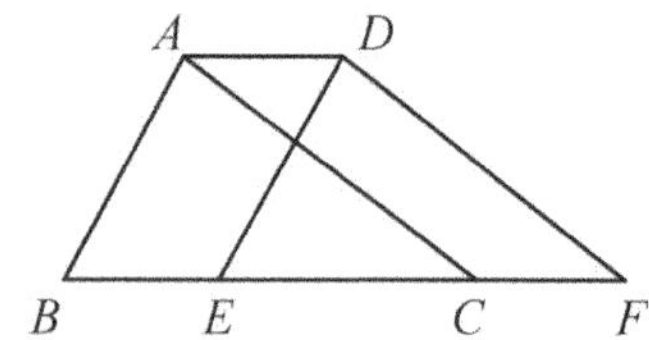

9 Look at the diagram. $\triangle ABC$ is folded along the line DE and the resulting figure is $\triangle A'DE$. $\angle A'EC = 32°$. Then $\angle A'ED =$ ________.

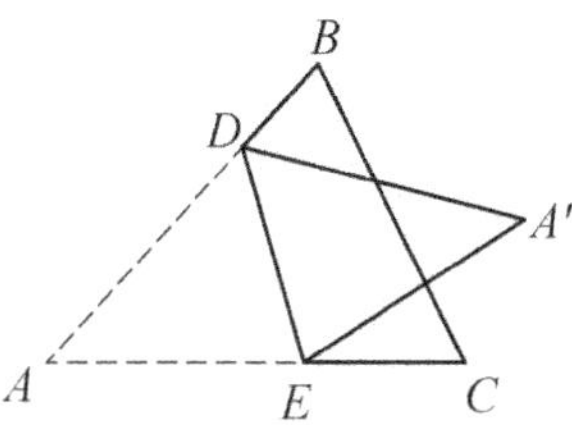

10 In the following diagram, triangle ABC has a right angle at vertex C, $AC = 5\,\text{cm}$, $BC = 12\,\text{cm}$ and $AB = 13$ cm. The triangle is rotated 90° clockwise about point A, and the resulting figure is $AB'C'$. In the course of the rotation, the total area covered by the right-angled triangle ABC is ________. (Give your answer in terms of π.)

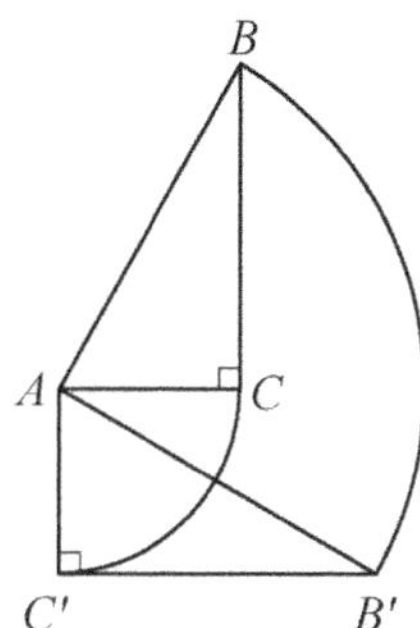

C. Questions that require solutions

11 On the coordinate plane below, first draw a triangle with vertices $A(0, 3)$, $B(2, 5)$ and $C(5, 2)$, and then apply the following transformations to the triangle:

(a) Reflect $\triangle ABC$ in the y-axis.

(b) First translate $\triangle ABC$ with vector $\begin{pmatrix} 0 \\ -3 \end{pmatrix}$, and then rotate the resulting figure 90° clockwise around the origin.

Compare the size, shape and position of the resulting figure in each of (a) and (b) with the original figure $\triangle ABC$. What has changed and what has not?

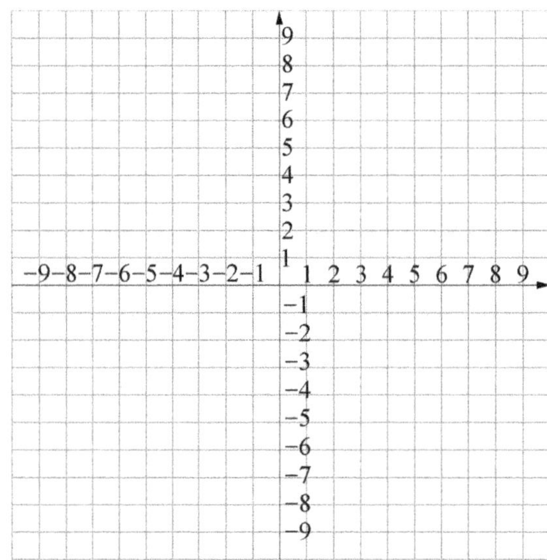

12 The diagram shows a square $ABCD$ with side length 1 cm. $DE = \frac{1}{4}$ cm and $AD = 1$ cm.

$\triangle ABF$ is obtained by rotating $\triangle ADE$.

(a) Write down the centre of the rotation, and the angle of the rotation.

(b) Find the area of $\triangle AEF$.

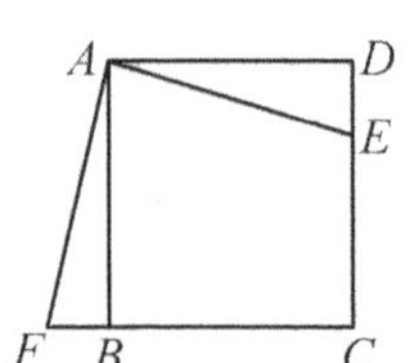

13 As shown in the diagram, point C is on line segment BE. An equilateral triangle ABC and an equilateral triangle DCE are constructed on the same side of BE. $\triangle ACE$ coincides with $\triangle BCD$ after a rotation.

(a) Write down the angle and the direction of the rotation.

(b) How many pairs of triangles, shown in the diagram, will coincide after a rotation?

(c) If angle 2 = 40°, then find $\angle BDE$.

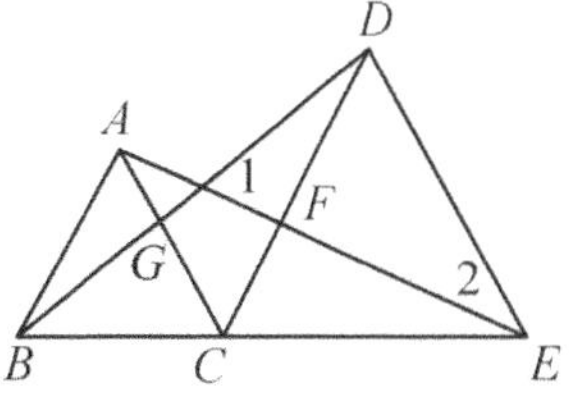

14 The diagram shows a triangular piece of paper ABC. The paper is folded so that point A coincides with point C, and the crease intersects the sides AC and BC at points D and E, respectively. The symmetric point of point B with respect to line DE is point F.

(a) Draw the line DE and point F.

(b) Join E to F and F to C. If $\angle FEC = 56°$, then find $\angle DEC$.

(c) Join A to E, B to D, and D to F. If $\frac{BE}{EC} = \frac{1}{3}$ and the area of $\triangle DEF$ is 2, then find the area of $\triangle ABC$.

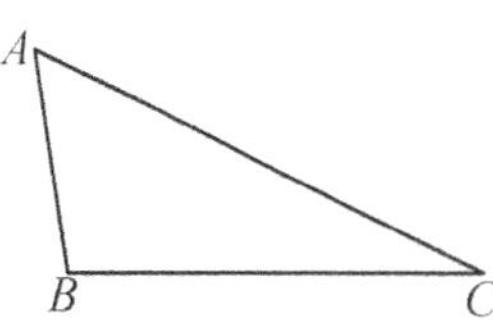

6.3 Equation of a circle

Learning objective

Recognise and use the equation of a circle and find the equation of a tangent to a circle at a given point.

A. Multiple choice questions

1 What is the equation of a circle with a radius of $\sqrt{5}$ and centre at the origin? (   )

A. $x^2 + y^2 = \sqrt{5}$
B. $x^2 - y^2 = \sqrt{5}$
C. $x^2 + y^2 = 5$
D. $x^2 - y^2 = 5$

2 What is the equation of a circle with a radius of 4 and centre at the point $(1, 3)$? (   )

A. $(x + 1)^2 + (y + 3)^2 = 16$
B. $(x - 1)^2 + (y + 3)^2 = 16$
C. $(x + 1)^2 + (y - 3)^2 = 16$
D. $(x - 1)^2 + (y - 3)^2 = 16$

3 Given that the equation of a circle is $x^2 + y^2 = 5$, then the equation of the tangent to the circle at point $(2, 1)$ is (   ).

A. $y = 2x$
B. $y = -2x + 5$
C. $y = 3x - 5$
D. $y = 2x - 3$

4 If the equation $4x^2 + 4y^2 - 8x + 4y - 3 = 0$ represents a circle, then the centre of this circle is (   ).

A. $\left(-1, -\frac{1}{2}\right)$
B. $\left(1, \frac{1}{2}\right)$
C. $\left(-1, \frac{1}{2}\right)$
D. $\left(1, -\frac{1}{2}\right)$

5 The equation of circle C_1 is $(x + 1)^2 + (y - 1)^2 = 1$. If circle C_2 is symmetrical to the circle C_1 with respect to the line $x - y - 1 = 0$, then the equation of circle C_2 is (   ).

A. $(x + 2)^2 + (y - 2)^2 = 1$
B. $(x - 2)^2 + (y + 2)^2 = 1$
C. $(x + 2)^2 + (y + 2)^2 = 1$
D. $(x - 2)^2 + (y - 2)^2 = 1$

B. Fill in the blanks

6 If the vertices of $\triangle OAB$ are $O(0, 0)$, $A(2, 0)$ and $B(0, 4)$, then the equation of the circumscribed circle of $\triangle OAB$ is ________.

7 Suppose that circle C with equation $(x-3)^2+(y-4)^2=1$ is symmetrical to circle M with respect to the x-axis. Then the equation of circle M is ________.

8 Suppose that line l passes through the point $(-1, 3)$, and is a tangent to the circle $x^2+y^2=10$. Then the slope of l is ________.

9 The centre of a circle is on the y-axis and its radius is 1. It passes through the point $(1, 3)$. Then its equation is ________.

10 Given that the equation of circle C is $(x-6)^2+(y+8)^2=4$ with centre at point C, and O is the origin, then the equation of the circle with OC as its diameter is ________.

C. Questions that require solutions

11 A circle passes through two points $A(1, 4)$ and $B(3, 2)$, and its centre is on the line $y=0$. Find the equation of the circle and determine the relationship between point $P(2, 4)$ and the circle.

12 The equation of a circle with centre O is $x^2+y^2=4$. Find the equation of the tangent to circle O at point $P(1, -\sqrt{3})$.

13 The centre of circle C is $C(k, k)$, and it passes through the point $P(6, 4)$.

(a) Write down the equation of circle C.

(b) What is the value of k when the area of circle C is the minimum? Find the equation of the circle C in this case.

14 A circle with radius 4 units has its centre in the first quadrant. Both the line $x = 0$ and the line $y = 1$, are tangents to the circle. Find the equation of the circle. (Hint: draw a sketch diagram to help you to work out the answer.)

6.4 Bearings

Learning objective

Interpret and solve problems using bearings.

A. Multiple choice questions

1. Which of the following statements about bearings is correct? (   )

   A. The bearing to a point is the angle measured clockwise from the east direction.

   B. The bearing to a point is the angle measured clockwise from the south direction.

   C. The bearing to a point is the angle measured clockwise from the north direction.

   D. The bearing to a point is the angle measured anti-clockwise from the north direction.

2. Look at the diagram. The bearing of point B from point A is (   ).

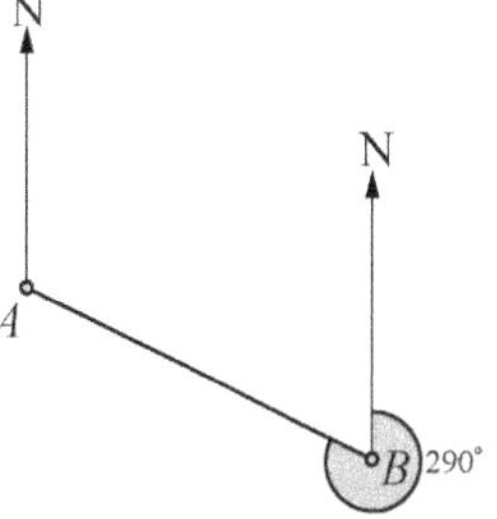

   A. 290°

   B. 070°

   C. 110°

   D. 250°

3. A student walks from point A on a bearing of 060° to point B, and then walks from point B on a bearing of 195° to point C. Then $\angle ABC$ is (   ).

   A. 30°   B. 45°

   C. 60°   D. 75°

4. Two boats M and N depart from port O at the same time. Boat M sails on a bearing of 330°, and boat N sails on a bearing of 250°. The two boats sail at the same speed. 1 hour after they leave port O, boats M and N arrive at points A and B, respectively. Then point B is at a bearing of (   ) from point A.

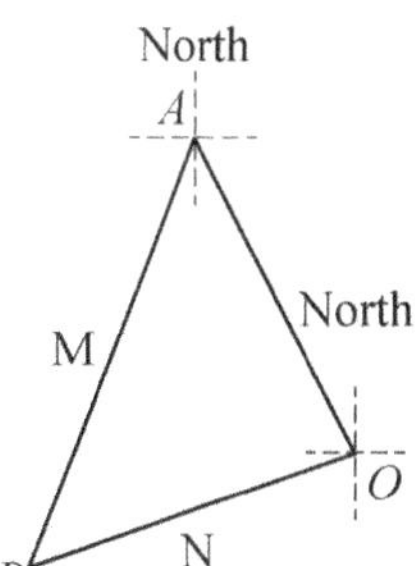

   A. 220°   B. 210°

   C. 200°   D. 190°

B. Fill in the blanks

5 Point B is 2 km from point A on a bearing of 030°, and point C is 2 km due east from point B. The exact distance between C and A is ________ km.

6 A ship sails due east. At 9 a. m. it is on a bearing of 190° from a lighthouse P and 120 nautical miles away from the lighthouse. At 11 a. m. it reaches point N which is due south of the lighthouse. During this time, the average speed of the ship is ________ nautical miles/hour.

7 A person stands at point C. In the distance, there is a pavilion B, and to the east of the pavilion B there is a tree A. Point A is on a bearing of ________ from C, and Point B is on a bearing of ________ from C.

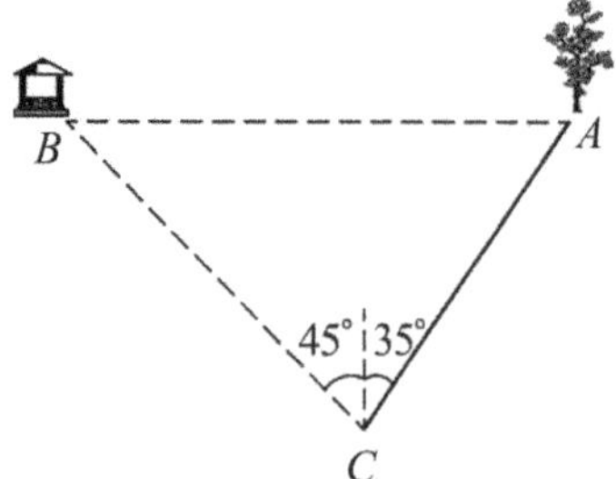

8 In Question 7, the distance between A and C is 100 metres. Then the distance between A and B is ________ metres (to 3 s. f.).

C. Questions that require solutions

9 Point P is 2 km from point Q on a bearing of 120°, and point M is 3 km from point P on a bearing of 210°.

(a) What is the bearing of M from Q?

(b) Find the distance between M and Q.

10 A motorcyclist sets off from home and drives due west for 30 km. He stops at a petrol station. He then drives to a shop which is another 30 km from the petrol station and on a bearing of 240°.

(a) At what bearing should he should set off to head directly home from the shop?

(b) What is the distance from the shop to his home?

11 Line l represents a straight sailing route. There are two observation points A and B, on each side of the route and the distance from point A to l is 2 km. Point B is on a bearing of 060° from point A, and it is 10 km away from A. A ship starts from point C which is on a bearing of 256° from point B, and sails along the route l from west to east. After 5 minutes, the ship reaches point D which is due north of point A.

(a) Find the distance from the observation point B to the route l.

(b) Find the speed of the ship (to the nearest 0.1 km/h).

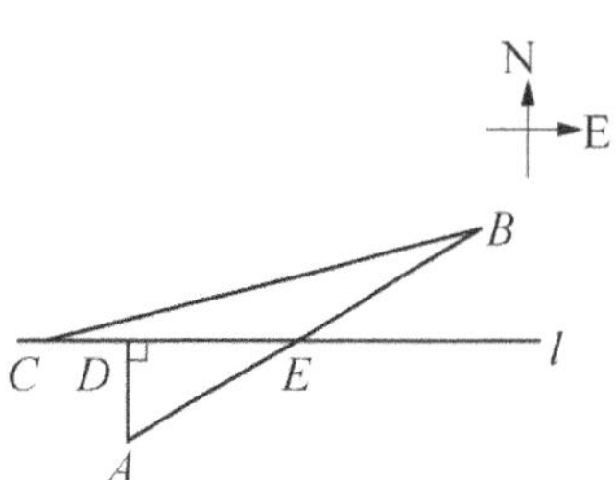

6.5 Introduction to plans and elevations of 3D shapes

Learning objective

Construct and interpret plans and elevations of 3D shapes.

A. Multiple choice questions

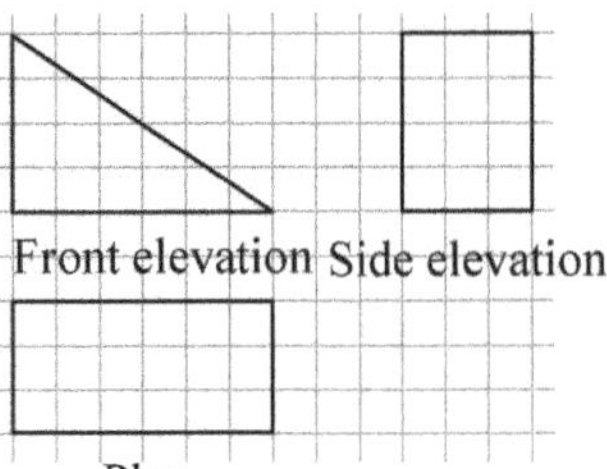

1. A 3D shape that would have a plan and elevation as shown in the diagram is a (　　).

   A. triangular pyramid
   B. triangular prism
   C. rectangular-based pyramid
   D. rectangular prism

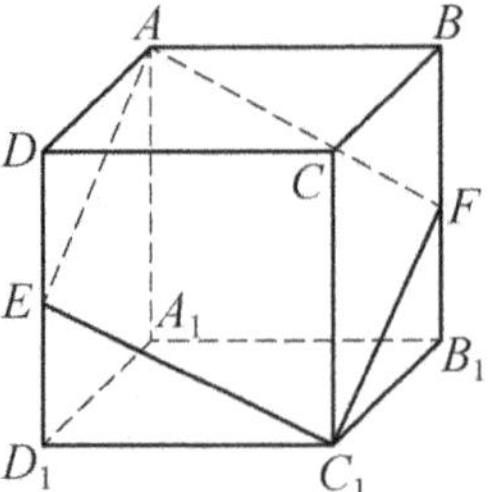

2. As shown in the diagram, in the cube $ABCDA_1B_1C_1D_1$, E and F are the midpoints of edges DD_1 and BB_1, and the upper part of the cube is cut by the plane which goes through A, E, C_1, F. Which of the following diagrams is the side elevation of the remaining solid (the lower part)? (　　)

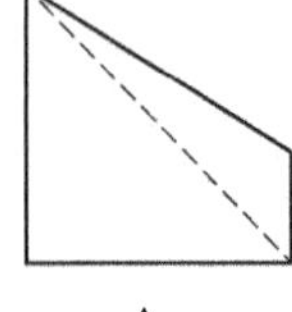
A.

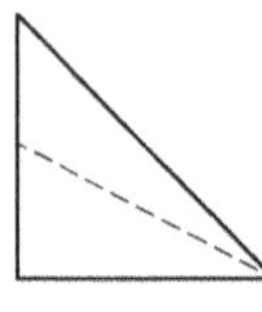
B.

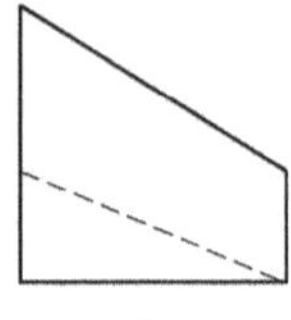
C.

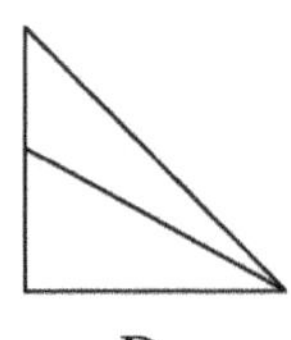
D.

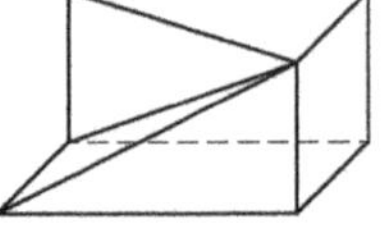

3. A rectangular-based pyramid is cut from a rectangular cuboid and the remaining part is shown in the diagram on the right. Then the front elevation of the remaining part is (　　).

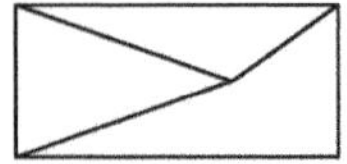
A.

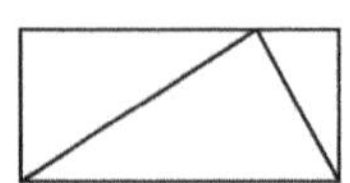
B.

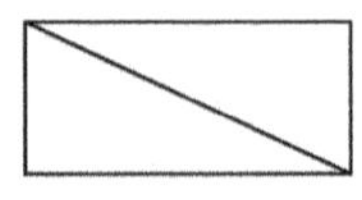
C.

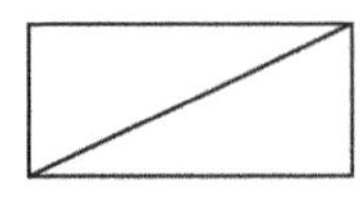
D.

4 A plan and elevation of a 3D shape is shown in the diagram. This 3D shape could be ( ).

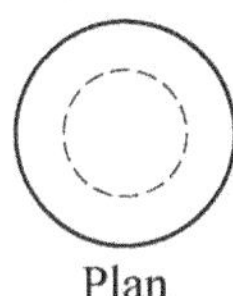

Plan

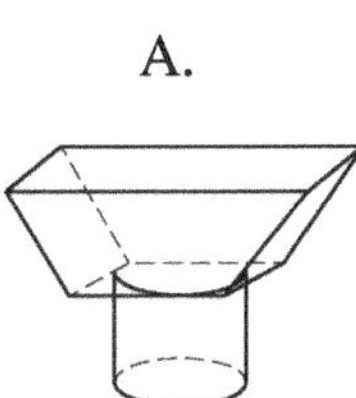

A.

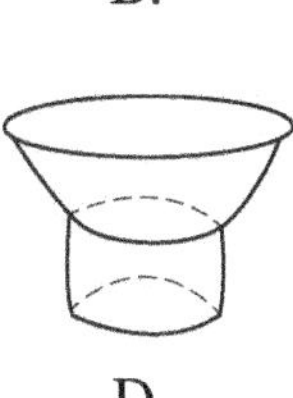

B.

C.

D.

B. Fill in the blanks

5 The front and side elevations of a 3D shape are as shown in the diagram. Then shape(s), ________ could be the plan of this 3D shape.

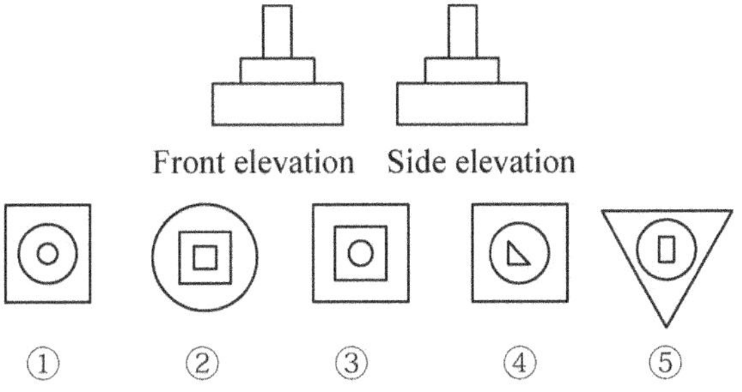

6 A square *EFGH* with side length 5 cm is an axial section of a cylinder (i. e., a cross-section passing through the cylindrical axis). The shortest distance from *E* to the opposite vertex *G* over the lateral surface is ________.

The **lateral surface** of a three-dimensional object is all of the sides of the object, excluding its bottom and top bases if they exist.

7 The diagram shows the front elevation and plan of a square-based pyramid. The slant height is ________.

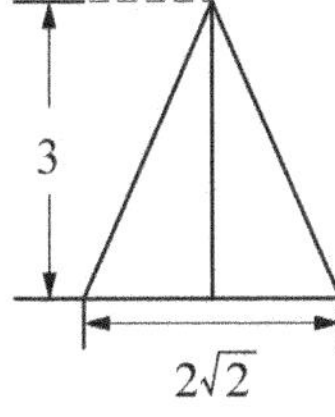
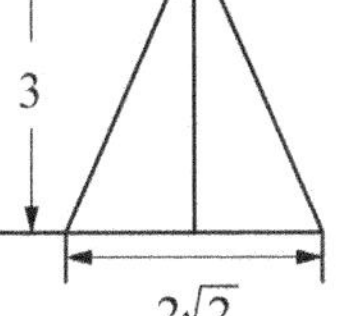

Front elevation

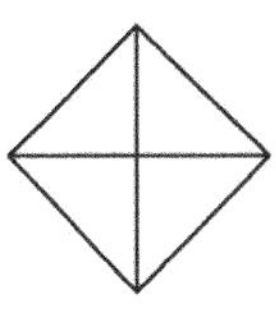

Plan

8 A part is cut off from a cylinder and then it is put together with a hemisphere of radius r to form a 3D shape with front elevation and plan as shown in the diagram. If the surface area of this 3D shape is $16 + 20\pi$, then $r =$ ________.

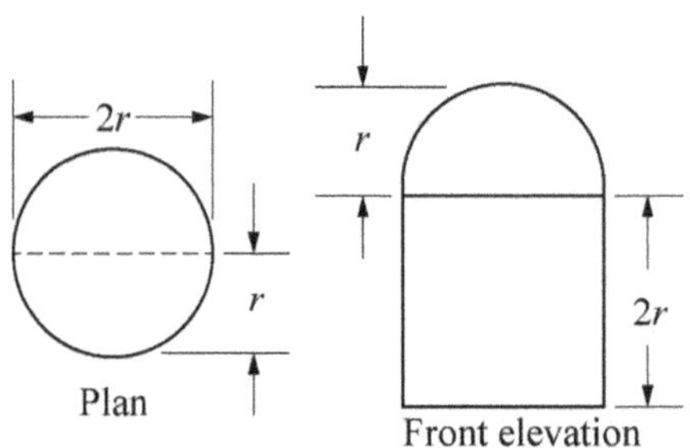

C. Questions that require solutions

9 The diagram shows a plan and elevation of a tetrahedron. Find the surface area of this tetrahedron.

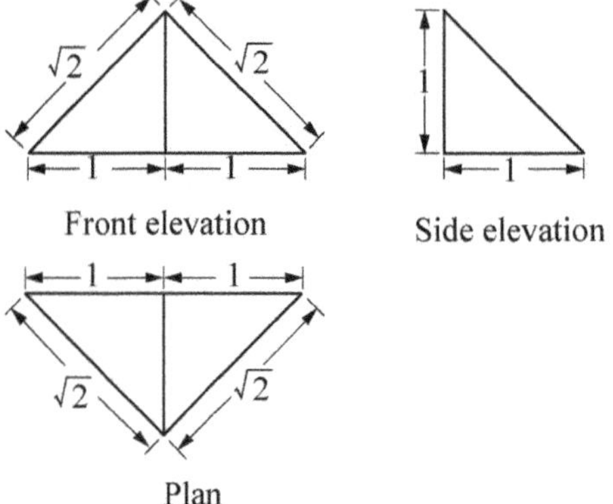

10 The diagram shows the plan, front elevation and side elevation of a solid 3D shape. Find the volume of the shape.

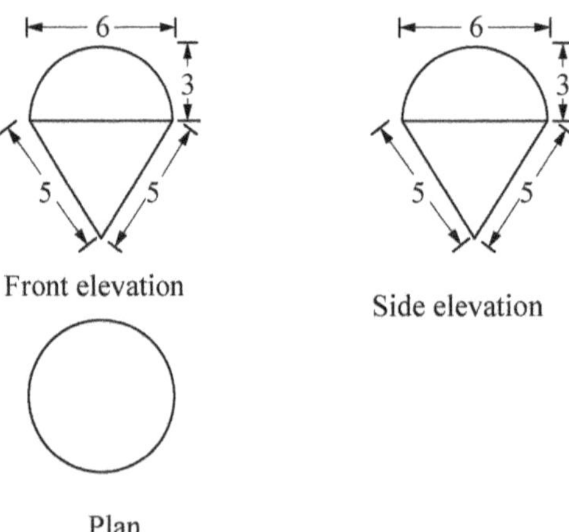

11 The height of a cylinder is 2, and the circumference of its base is 16. The front elevation, left-side elevation and the plan of this cylinder are as shown in the diagram. A point M on the surface of the cylinder corresponds to the point A on the front elevation, and a point N on the surface of the cylinder corresponds to the point B on the left-side elevation. Then, on the lateral surface of this cylinder, for all routes starting from M and ending at N, the length of the shortest route is ________.

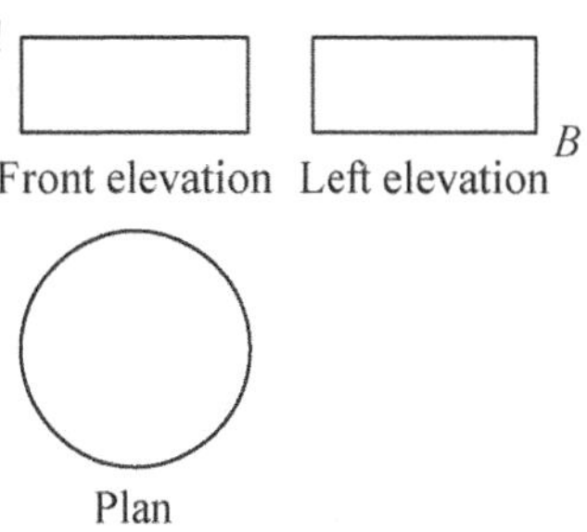

12 The diagram shows a cone. Given that the radius of the base of the cone is $r = 1$, the slant $l = 4$, M is a point on the slant SA, and $SM = x$. A string is fixed at point M, and is then wrapped around the lateral surface until it reaches the point A.

Find:

(a) the square of the shortest length of this string, denoted as $f(x)$.

(b) the shortest distance from the apex to any point of the string when the string is shortest (in terms of x).

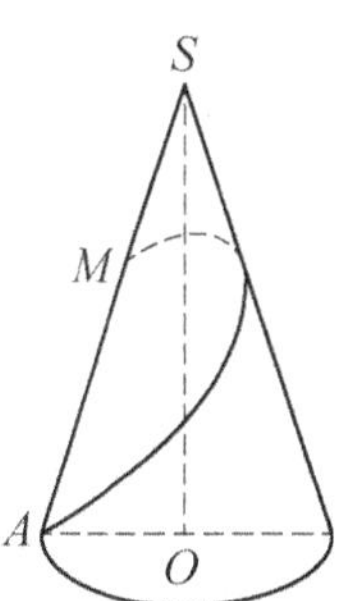

6.6 Surface areas of 3D shapes

Learning objective

Calculate and solve problems involving the surface area of 3D shapes.

A. Multiple choice questions

1 Which of the following statements is correct? (　　).

A. All geometric shapes have surface areas.

B. There are no surface areas for some 3D shapes.

C. The surface area of a 3D shape measures how much area the surfaces of that shape have in total.

D. The surface area of a 3D shape measures how much space that shape occupies.

2 A sphere is cut into two equal halves. The surface area of each half is (　　) half of the surface area of the sphere.

A. equal to　　B. larger than

C. less than　　D. uncertain

3 When two identical cubes are put together to form a cuboid. The surface area of the cuboid is (　　) twice the surface area of the two separate cubes.

A. equal to　　B. larger than　　C. less than　　D. uncertain

B. Fill in the blanks

4 If the height of a cone is 8 cm and the radius of its base is 6 cm, then the surface area of the cone is ________ cm^2.

5 The area of the lateral surface of a cone is twice the area of its base. The sector part of the net of the cone has a central angle of ________ degrees.

6 The lateral surface of a cone is made of a piece of iron sheet in the shape of a sector with radius 9 cm and central angle 240°. The base of the cone is made of a circular iron disc. The surface area of this cone is ________ cm^2.

7 A sphere is cut by a plane, whose distance to the centre of the sphere is 1, and the area of the circle obtained from the cutting is π. The surface area of the sphere is ________.

8 Given that the radii of three spheres R_1, R_2 and R_3 satisfy $R_1 + 2R_2 = 3R_3$, then an equation that their surface areas of S_1, S_2 and S_3 satisfy is ________.

9 The radius of a sphere with centre O is $OA = R$. A plane goes through the midpoint M of OA and is perpendicular to OA. The plane cuts the sphere. The section obtained is a circle with centre M. Then the ratio of the area of the circle to the surface area of the sphere is ________.

C. Questions that require solutions

10 In the triangular-based pyramid $SABC$ shown in the diagram, $\angle SAB = \angle SAC = \angle ACB = 90°$, $AC = BC = 5$ cm and $SB = 5\sqrt{5}$ cm. Find the surface area of this pyramid.

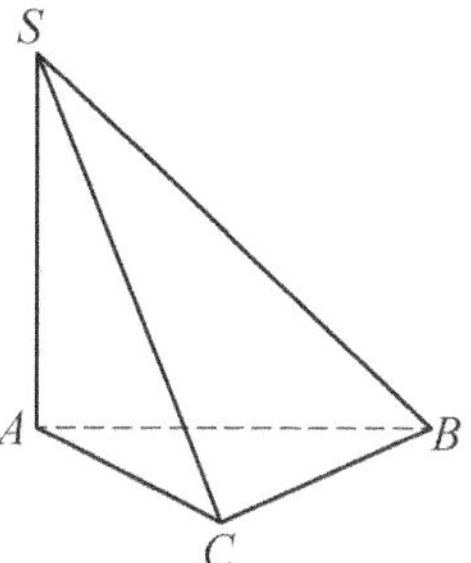

11 In $\triangle ABC$, $\angle C = 90°$, $AC = 20$ cm and $BC = 15$ cm. This right-angled triangle is rotated 360° about AB. Find the surface area of the solid obtained by the rotation.

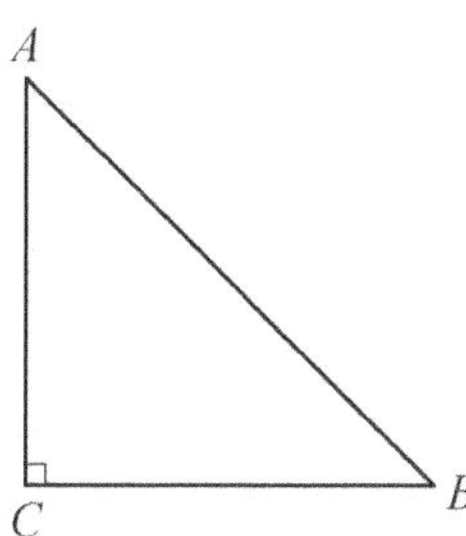

12 The base diameter and height of a cone are both equal to the diameter of a sphere. Find:

(a) the ratio of the surface area of the sphere to the lateral area of the cone

(b) the ratio of the surface area of the sphere to the surface area of the cone.

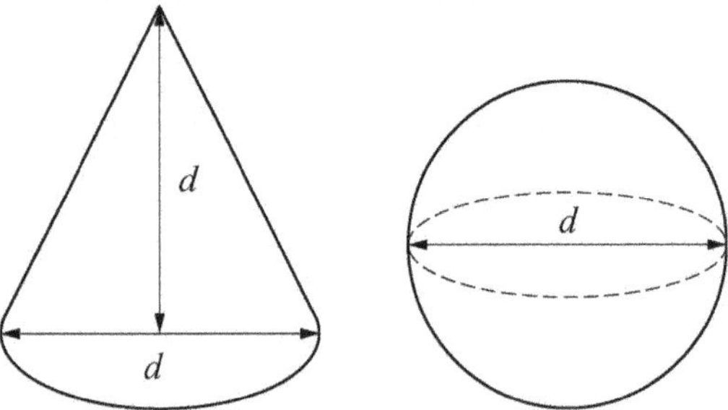

6.7 Volumes of 3D shapes

Learning objective

Calculate and solve problems involving the volume of 3D shapes.

A. Multiple choice questions

1 Complete this statement: The volume of a cube with edge length 1 is (   ) the volume of a sphere with radius 1.

A. equal to　　B. larger than

C. less than　　D. uncertain

2 If three identical cubes are put together to form a cuboid, which of the following statements is correct? (   )

A. The surface area of the cuboid is triple the surface area of each cube.

B. The volume of the cuboid is triple the volume of each cube.

C. The surface area of the cuboid is larger than triple the surface area of each cube.

D. The volume of the cuboid is larger than triple the volume of each cube.

3 Which of the following statements is/are correct? (   )

A. If two 3D shapes have the same surface area, then they also have the same volume.

B. If two 3D shapes have the same volume, then they also have the surface area.

C. For all 3D shapes, the larger the surface area, the larger the volume.

D. For all spheres, the larger the surface area, the larger the volume.

B. Fill in the blanks

4 The square-based pyramid shown in the diagram has base length 6 cm and slant length 5 cm. Its volume is ________.

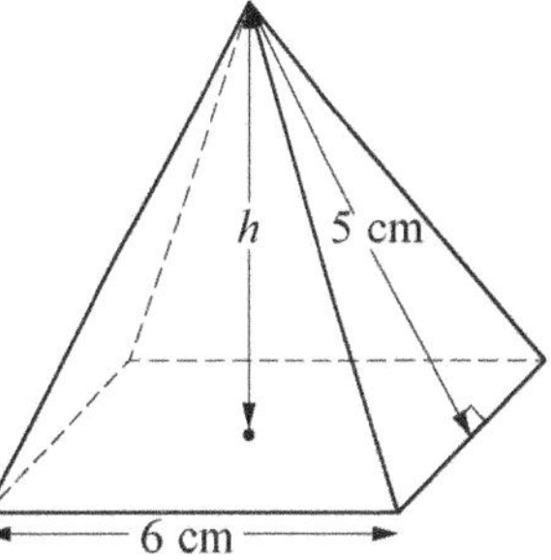

5 If the surface area of a sphere is 16π cm^2, then the volume of this sphere is ________ cm^3.

6 The radii of three balls are R_1, R_2 and R_3, satisfying $R_1 + R_2 = 3R_3$ then the equation that their volumes V_1, V_2, and V_3 satisfy is ________.

7 The surface areas of a sphere and a cube are equal. Then the ratio of the volume of the sphere to the volume of the cube is ________.

8 A composite solid has 9 faces, and all its edges have length 1 cm. A net of the solid is shown in the diagram. The volume of the convex polyhedron formed from this net is ________.

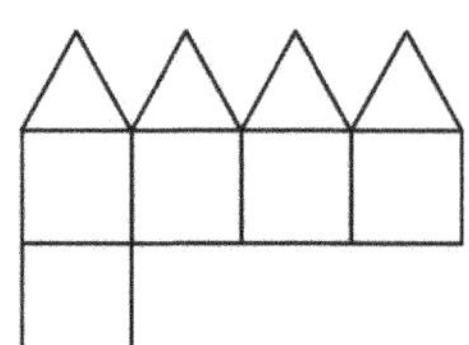

C. Questions that require solutions

9 An empty cone-shaped cup has an ice-cream sphere on top of it. Both the diameter of the base of this cone and the diameter of the sphere are 10 cm. All of the ice cream melts and flows down into the empty cone-shaped cup which it fills without any spilling. What is the minimum height of the cup?

10 The edge length of a cube is 2 cm. Find the volume of the solid whose vertices are all the centres of the faces of this cube, as shown in the diagram.

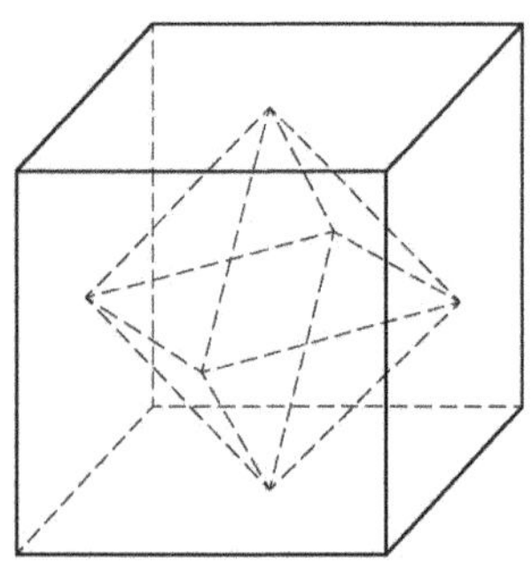

11 The slant height of a cone is l, and the angle between a slant and the base is θ. The four vertices of the top face of a cube are on a slanting edge of the cone, and the four vertices of the bottom face are on the base of the cone. Find the ratio of the volume of the cube to the volume of the cone.

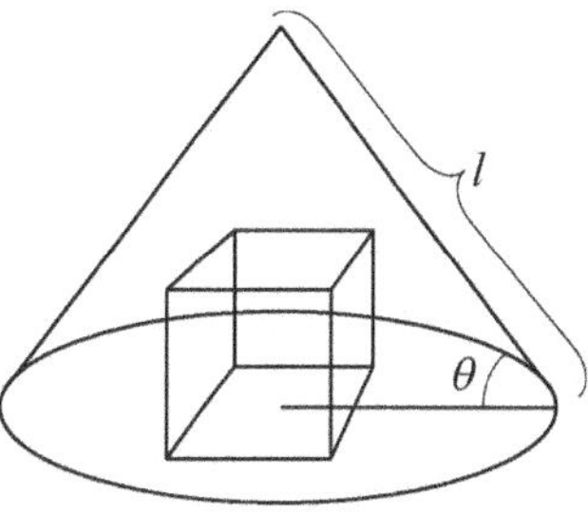

12 The diagram shows a cone for which the radius of the base is 2 cm and the height is 6 cm. There is an inscribed cylinder whose height is x inside the cone.

(a) Write an expression for the curved surface area of the cylinder in terms of x.

(b) What value of x gives the maximum curved surface area of the cylinder?

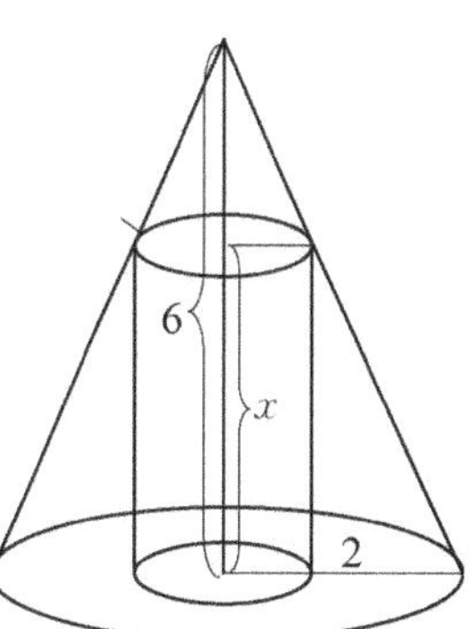

6.8 Measurement involving congruent and similar figures (1)

Learning objective

Use and apply the properties of congruence and similarity, including the relationships between lengths, areas, and volumes in similar figures.

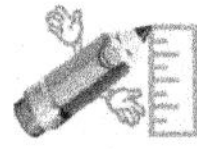

A. Multiple choice questions

1 Which of the three rectangles shown in the diagram are similar? (   )

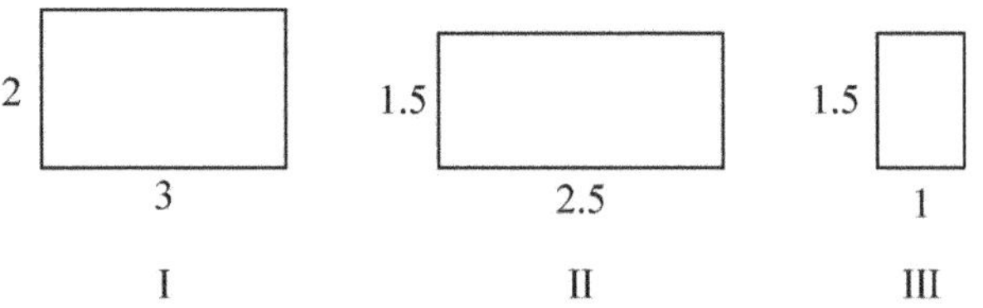

A. I and II B. I and III C. II and III D. none of them

2 If the perimeter of $\triangle ABC$ is 20 cm, and the points D, E and F are the midpoints of the three sides of $\triangle ABC$, then the perimeter of $\triangle DEF$ is (   ) cm.

A. 5 B. 10 C. 15 D. $\frac{20}{3}$

3 The diagram shows a right-angled triangle ABC with three squares drawn inside it. The side lengths of the squares, a, b and c form a geometric sequence. The relationship between a, b and c is (   ).

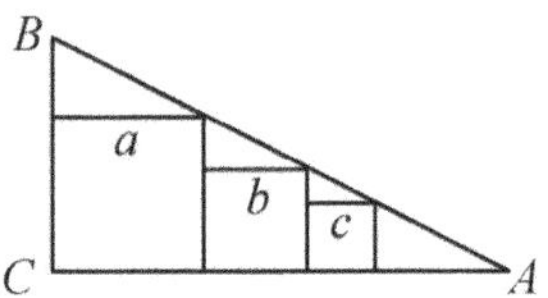

A. $b = a + c$ B. $b^2 = ac$

C. $b^2 = a^2 + c^2$ D. $b = 2a = 2c$

4 Ikenna uses a 3.2-metre-long bamboo pole to measure the height of a flagpole. He places the bamboo pole so that the shadows of both the top of the bamboo pole and the top of the flagpole coincide at a point on the ground. The distance from the bamboo pole to the point where the

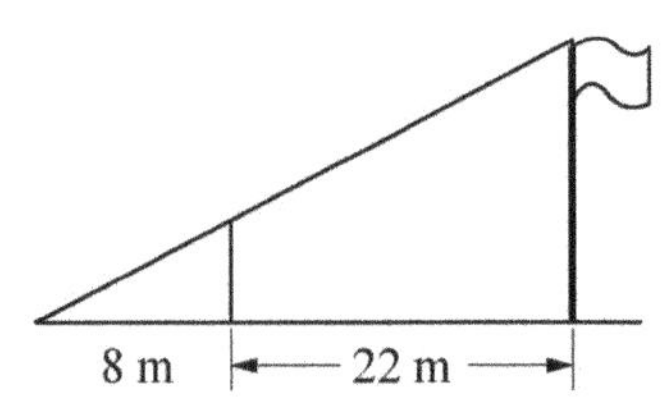

shadows coincide is 8 metres and the distance between the bamboo pole and the foot of the flagpole is 22 metres, as shown in the diagram. The height of the flagpole is ( ).

A. 12 m B. 10 m C. 8 m D. 7 m

B. Fill in the blanks

5 The difference between the areas of two similar triangles is $9\,\text{cm}^2$, and the ratio of their corresponding heights is $\sqrt{2} : \sqrt{3}$. The areas of these two triangles are ________ and ________.

6 The diagram shows a trapezium $ABCD$. AD is parallel to BC, and AC and BD intersect at point O. If $AD : BC = 3 : 7$, then $AO : OC =$ ________, and the ratio of the areas of the triangles $\text{area}_{\triangle AOD} : \text{area}_{\triangle BOC} =$ ________.

7 If the radius of a sphere is doubled, then its surface area is increased by ________ times.

8 Given that the ratio of the surface areas of two cubes is 3, then the ratio of their edge lengths is ________.

9 A cone is converted into a new cone. If the radius of the base of the cone is reduced by half, and the volume remains the same, then the height is increased by ________ times.

C. Questions that require solutions

10 Given that the side lengths of $\triangle ABC$ are 5, 12 and 13, and the length of the longest side of a similar triangle $A'B'C'$ is 26, find the area of $\triangle A'B'C'$.

11 As shown in the diagram, $\triangle ADE$ is similar to $\triangle ABC$. DE intersects AB at point D and intersects AC at point E. $DE = 2\,\text{cm}$, $BC = 5\,\text{cm}$ and the area of $\triangle ABC = 20\,\text{cm}^2$. Find the area of $\triangle ADE$.

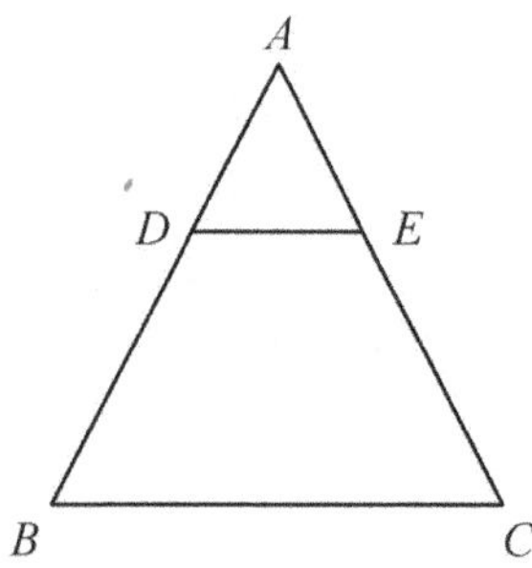

12 The diagram shows $\triangle ABC$ in which $AB = 14\,\text{cm}$, $\frac{AD}{BD} = \frac{5}{9}$, DE is parallel to BC, CD is perpendicular to AB and $CD = 12\,\text{cm}$. Find the area and perimeter of $\triangle ADE$.

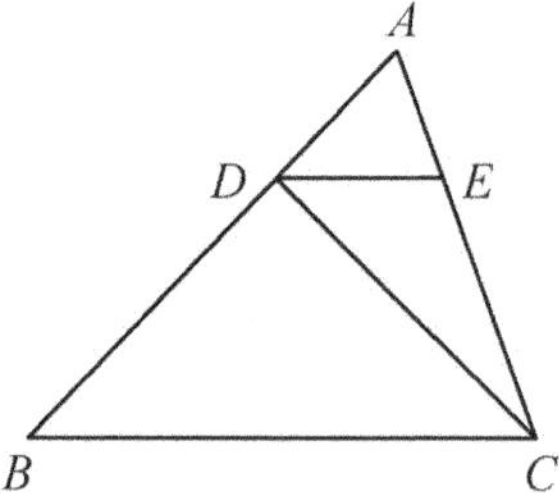

13 A plane is parallel to the base of a square-based pyramid, and it passes through the midpoint of the height from the apex to the base, as shown in the diagram.

(a) What is the ratio of the area of cross section to the area of the base?

(b) If the base is not a square, will the ratio be changed?

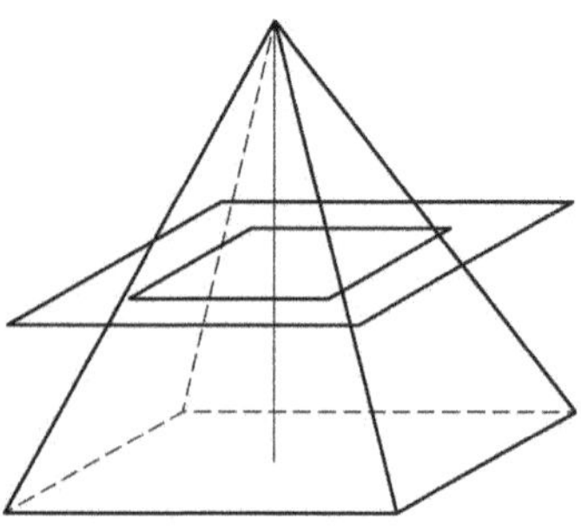

6.9 Measurement involving congruent and similar figures (2)

Learning objective

Use scale factors for length, area and volume in ratio form to solve problems involving similar figures.

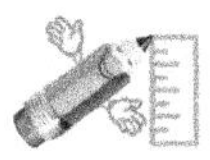

A. Multiple choice questions

1 If the ratio of the lengths of the bisectors of a pair of corresponding angles of two similar triangles is 16 : 25, then the ratio of the areas of the triangles is (   ).

A. 4 : 5
B. 16 : 25
C. 196 : 225
D. 256 : 625

2 A pyramid is cut by a plane that is parallel to its base. If the ratio of the area of the cross section to the area of the base is 1 : 2, then the height of the pyramid is divided into two segments (upper and lower) in a ratio of (   ).

A. $1 : \sqrt{2}$
B. 1 : 4
C. $1 : (\sqrt{2} + 1)$
D. $1 : (\sqrt{2} - 1)$

3 Two cross sections, both parallel to the base of a pyramid, cut the pyramid through the two points on its height which divide the height into three equal parts. The ratio of the volumes of the three parts obtained is (   ).

A. 1 : 1 : 1
B. 1 : 2 : 3
C. 1 : 8 : 27
D. 1 : 7 : 19

4 If the surface area of a sphere is increased so that the new surface area is n times as large as the original one, then the new radius is (   ) times as long as the original radius.

A. $\sqrt{n} - 1$
B. $\sqrt{n}$
C. $\sqrt{n} + 1$
D. $\sqrt{n} + 2$

B. Fill in the blanks

5 The ratio of the areas of two similar triangles is 9 : 16. If the perimeter of the smaller triangle is 36 cm, then the perimeter of the larger triangle is ________.

6 As shown in the diagram, M is a point inside $\triangle ABC$. Three lines are constructed through M so that they are parallel to the three sides of $\triangle ABC$. If the areas of the three triangles formed, i.e., $\triangle_1$, $\triangle_2$ and $\triangle_3$ (as shown by the shaded areas in the diagram), are $4\ \text{cm}^2$, $9\ \text{cm}^2$ and $49\ \text{cm}^2$, respectively, then the area of the $\triangle ABC$ is ________.

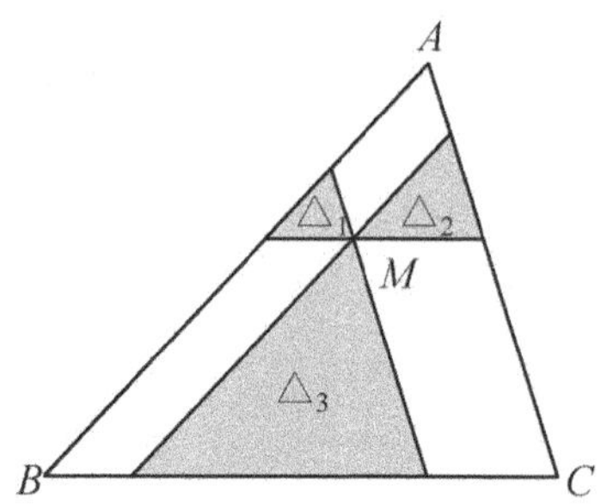

7 The diagram shows $\triangle ABC$. Points D and E are on AB and AC, respectively, and $\angle ADE = \angle C$. If $AE = 2\ \text{cm}$, the area of $\triangle ADE$ is $4\ \text{cm}^2$ and the area of the quadrilateral $BCED$ is $5\ \text{cm}^2$, then the length of AB is ________.

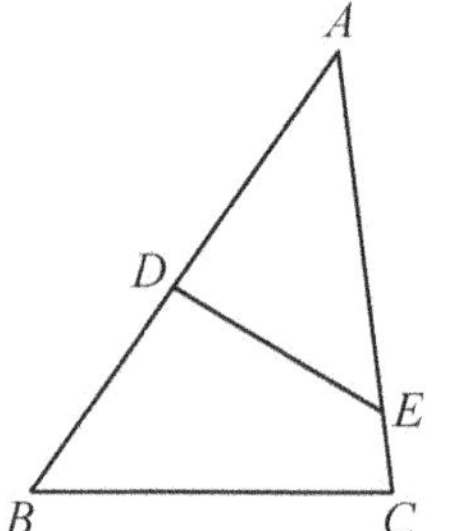

8 There are two cones A and B. Given that the radius of the base of cone A is half that of cone B, the slant height of cone A is twice that of cone B, and the height of cone A is $\sqrt{\frac{37}{8}}$ times that of cone B, then the ratio of the surface area of cone A to the surface area of cone B is ________.

9 A small pyramid is cut from a larger pyramid by making a cut parallel to the base. The ratio of the base area of the small pyramid to the base area of the larger pyramid is $\frac{1}{9}$. The ratio of the volume of the small pyramid to the volume of the larger pyramid is ________.

C. Questions that require solutions

10 In $\triangle ABC$, $AB = 2\sqrt{3}$ and $AC = 2$. AD is the height perpendicular to the side BC and $AD = \sqrt{3}$.

(a) Find the length of BC.

(b) A square can be drawn with one side on AB and the other two vertices on AC and BC, respectively. Find the area of the square.

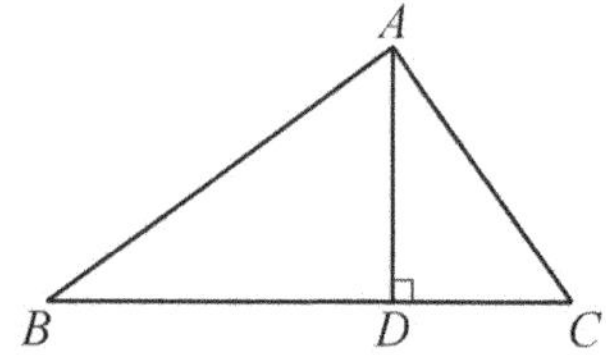

11 The lengths of a pair of corresponding sides of two similar triangles are 24 cm and 12 cm.

(a) If the sum of their perimeters is 120 cm, find the perimeters of these two triangles.

(b) If the difference in their areas is $420\ \text{cm}^2$, find the areas of these two triangles.

12 The area of the base of a cone is $150\ \text{cm}^2$, the area of a cross section parallel to the base is $54\ \text{cm}^2$ and the distance between the cross section and the base is 14 cm. What is the height of this cone?

13 A regular triangular pyramid is a pyramid whose base is an equilateral triangle and whose lateral edges are all equal in length. The diagram shows a regular triangular pyramid $SABC$. Given that its height is $SO = h$ and the slant height is $SM = l$, find the area of the cross section $A_1B_1C_1$ that passes through the midpoint of SO and is parallel to the base.

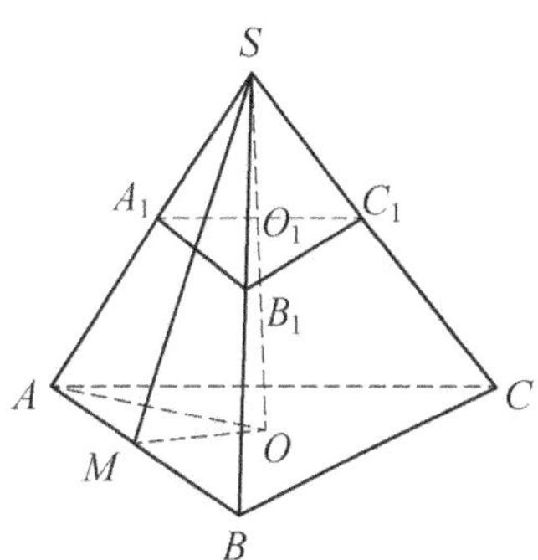

Unit test 6

A. Multiple choice questions

1 A rectangle is 12 cm long and 6 cm wide. After an enlargement, it is 36 cm long and 18 cm wide. The scale factor of the enlargement is (　　).

A. 9　　B. 1

C. 3　　D. 4

2 A plan and elevation of a 3D shape are shown in the diagram. This 3D shape could be (　　).

Front elevation

Side elevation

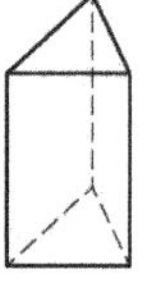

A.

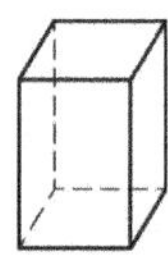

B.

C.

D.

Plan

3 A rectangular piece of paper is folded as shown in the diagram. Given that $\angle\alpha = 36°$, then $\angle\beta =$ (　　).

A. 36°

B. 72°

C. 60°

D. 45°

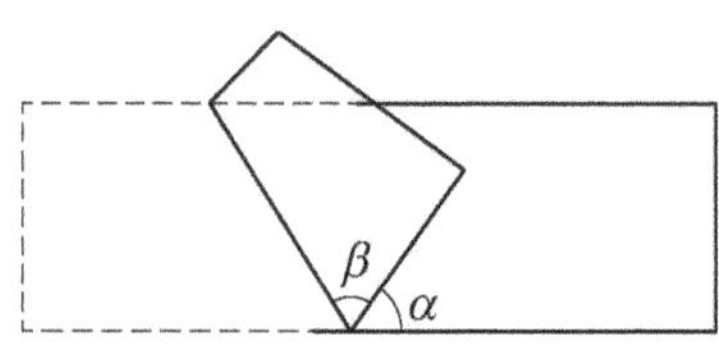

4 The perimeter ratio of two similar triangles is 2 : 3. If the area of the larger triangle is 27 cm^2, then the area of the smaller triangle is (　　).

A. 12 cm^2　　B. 18 cm^2

C. 24 cm^2　　D. 27 cm^2

5 As shown in the diagram, triangle $A'B'O$ is an enlargement of triangle ABO with a scale factor of -3. Given that the coordinates of point B are $(-2, 3)$, the coordinates of B' are ( ).

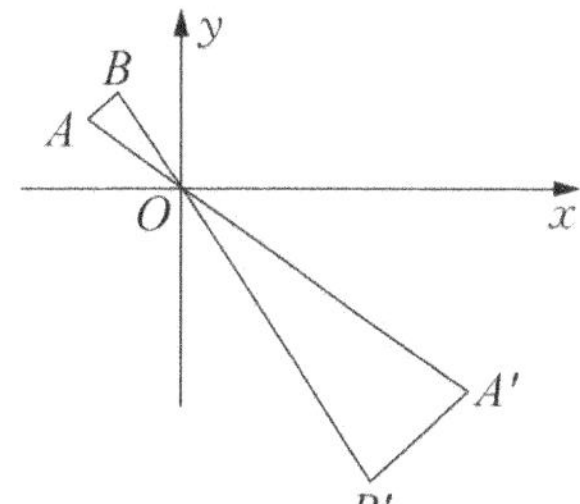

A. (6, −9)
B. (9, −6)
C. (6, −4)
D. (4, −6)

6 The centre of a circle is on the y-axis. Given that the radius is 5 and the circle passes through the point $(-3, 4)$, then the equation of the circle could be ( ).

A. $x^2 + (y - 8)^2 = 25$
B. $x^2 + (y + 8)^2 = 25$
C. $(x + 3)^2 + (y - 4)^2 = 25$
D. $(x - 8)^2 + y^2 = 25$

7 Island C is on a bearing of 105° from point A and point B is on a bearing of 150° from A. A boat travels from A for 30 minutes to point B at a constant speed of 60 mph. The island C is north-east of B. The distance from B to the island C is ( ) miles.

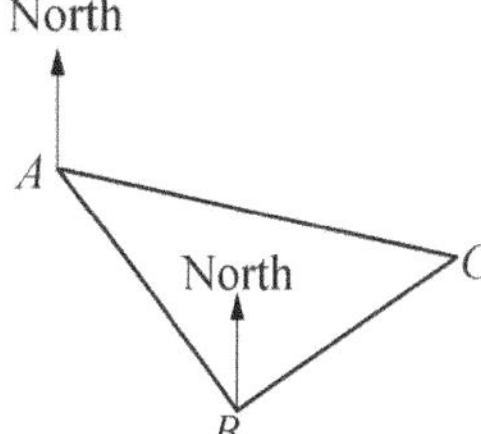

A. $10\sqrt{2}$
B. 60
C. $10\sqrt{6}$
D. 30

8 The diagram shows a cylinder that has been cut and put together to form an approximate cuboid. Compare the cuboid with the original cylinder. Which of the following statements is correct? ( )

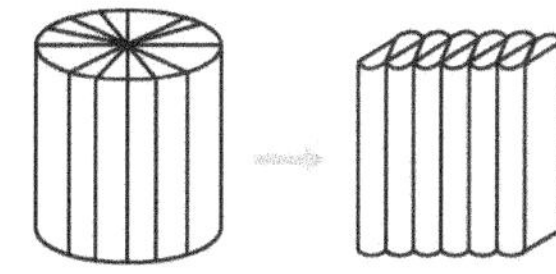

A. The surface area has not changed, but the volume has.
B. The surface area has changed, but the volume has not.
C. The surface area and volume have both changed.
D. The surface area and volume have both not changed.

9 As shown in the diagram, triangle ABC is enlarged to form triangle DEF. The centre of the enlargement is point O, and the ratio of the area of ABC to the area of DEF is 4 : 9. The ratio of lengths AO : OD is ( ).

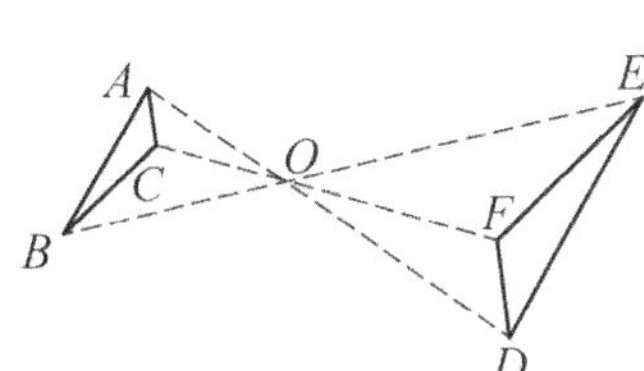

A. 2 : 5
B. 4 : 9
C. 4 : 13
D. 2 : 3

B. Fill in the blanks

10 A triangle ABC is enlarged to form a similar triangle $A'B'C'$, as shown in the diagram. Given that $AB = 2.4$, $A'B' = 3.6$, $BC = 1$ and $A'C' = 2.7$, then $B'C' =$ ________ and $AC =$ ________.

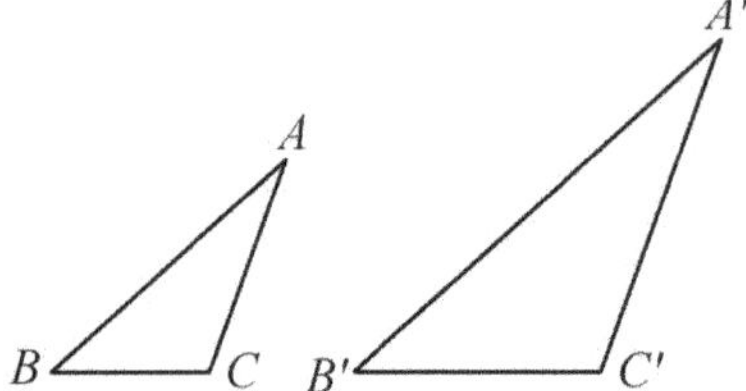

11 Two sides of a right-angled triangle ABC are $AC = 8$ cm and $BC = 6$ cm. Angle $ACB = 90°$. This right-angled triangle is rotated 360° about line AC. The surface area of the solid obtained after the rotation is ________ cm^2.

12 The diagram shows a plan and elevation of a 3D shape. Its surface area is ________.

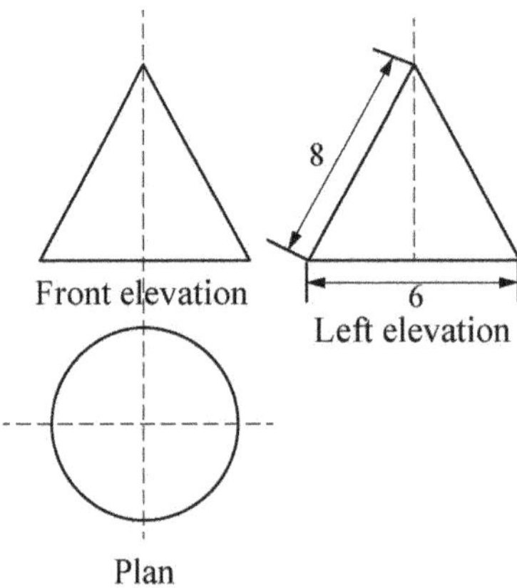

13 The actual length of a scenic railway is about 26 kilometres. On a map with scale 1 : 800 000, the length of the scenic railway is ________ centimetres.

14 As shown in the diagram, triangle ABC and triangle DEF are similar. The ratio of corresponding lengths in the two triangles is 2 : 3. If the perimeter of triangle ABC is 4 cm, then the perimeter of triangle DEF is ________.

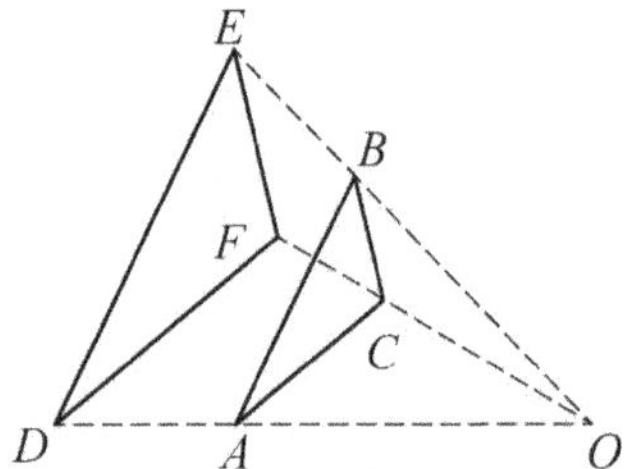

15 If the sum of the lengths of all the edges of two cubes is 48 cm and the sum of the surface areas is $72\ cm^2$, then the sum of the volumes is ________ cm^3.

16 The side length of the square $ABCD$ shown in the diagram is 3 cm. E is a point on AB, $AE = 1$, and $\triangle DCF$ is obtained after $\triangle DAE$ is rotated 90° anti-clockwise around point D. The length of EF is ________.

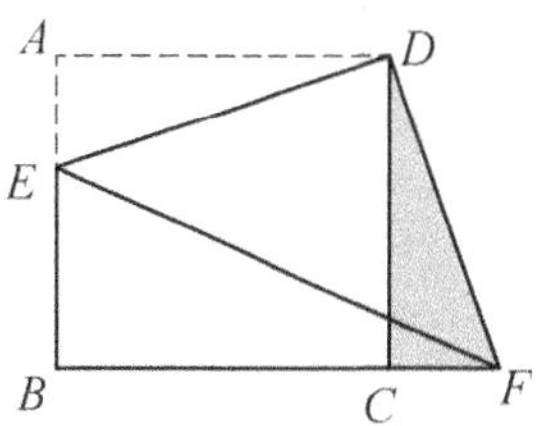

C. Questions that require solutions

17 The diagram shows the graph of a linear function.

(a) Write the equation of the linear function.

(b) Write the equation of the circle with A as the centre and AB as the radius.

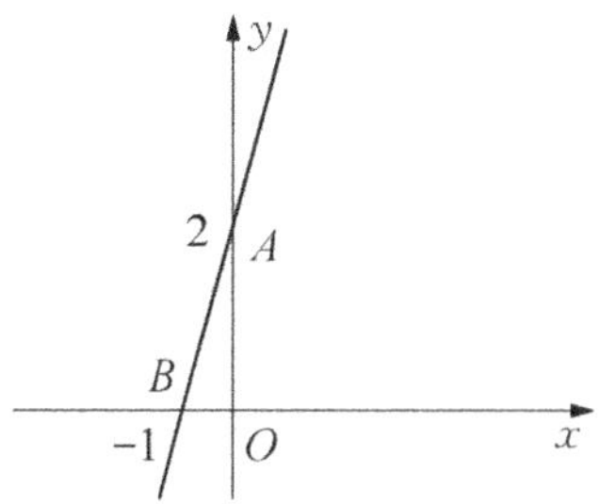

18 As shown in the diagram, point O is the origin of a coordinate plane, B has coordinates $(3, -1)$, and C has coordinates $(2, 1)$.

(a) Draw $\triangle O'B'C$ which is an enlargement of $\triangle OBC$ with a scale factor of -2, centre C.

(b) Find the area of $\triangle O'B'C$.

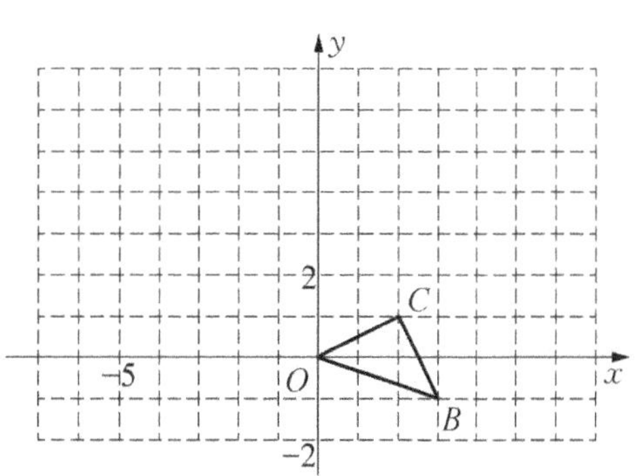

19 $\triangle ABE$ is similar to $\triangle ADB$. E is a point on AD, $\frac{BE}{DB} = \frac{3}{5}$, $\angle AEB = 110°$ and $\angle A = 40°$.

(a) Find the sizes of $\angle ABD$ and $\angle D$.

(b) Find the ratio of corresponding sides of $\triangle ABE$ and $\triangle ADB$.

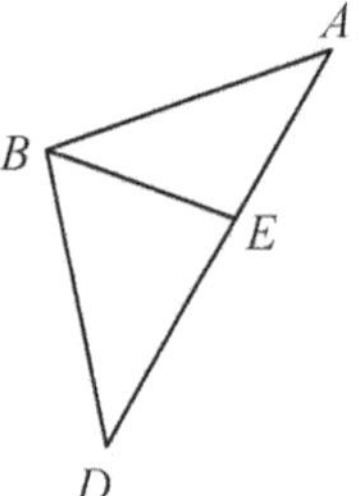

20 In $\triangle ABC$, $AC > AB$ and point D is on AC (not coincident with A and C), as shown in the diagram.

(a) State another condition that makes $\triangle ABD$ similar to $\triangle ACB$.

(b) Using the condition you gave in (a), if $AD : DC = 1 : 2$, find the ratio of corresponding lengths and the ratio of the area of $\triangle ABD$ to the area of $\triangle ACB$.

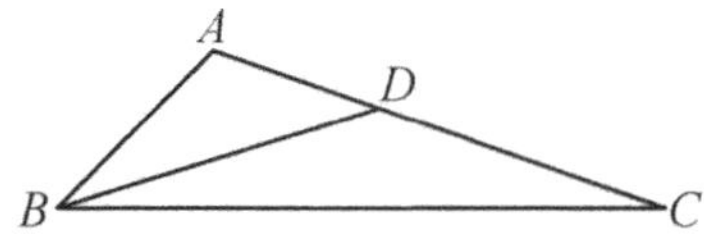

21 Islet B is south east of deep-water port A. A ship departs from port A and travels at a constant speed of 30 kilometres per hour due east. After 40 minutes, it arrives at C from which islet B is on a bearing of 165°. Find the distance from islet B to port A. (Give your answer correct to 1 decimal place.)

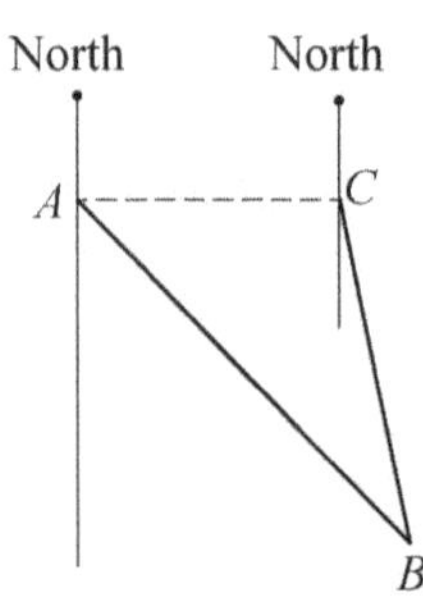

Chapter 7 Probability

7.1 Systematic listing and the product rule

Learning objective

Enumerate sets and combinations of sets systematically.

A. Multiple choice questions

1 If two different natural numbers from 1 to 10 inclusive are selected so that their sum is not greater than 10, then there are ( ) possible selections.

A. 10 B. 20 C. 30 D. 40

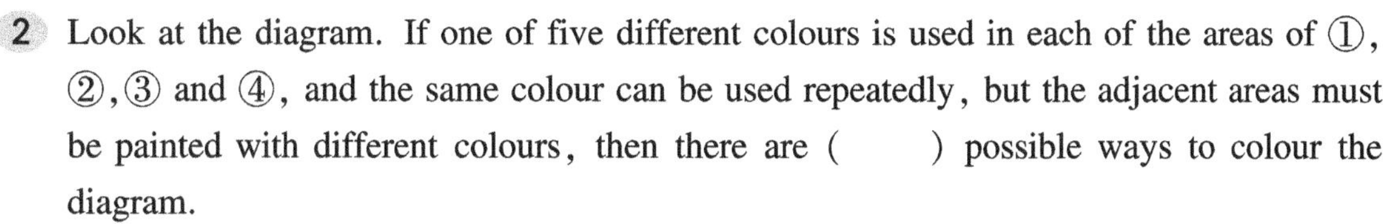

2 Look at the diagram. If one of five different colours is used in each of the areas of ①, ②, ③ and ④, and the same colour can be used repeatedly, but the adjacent areas must be painted with different colours, then there are ( ) possible ways to colour the diagram.

A. 180
B. 160
C. 96
D. 60

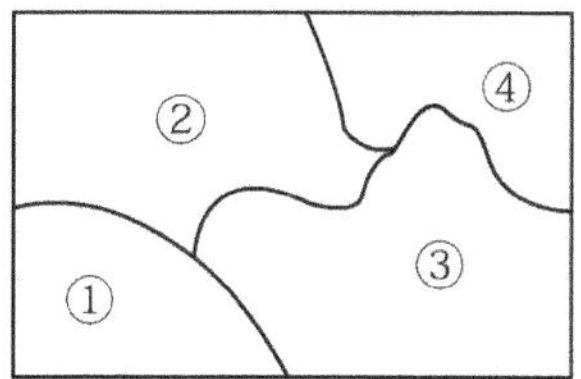

3 If the letters a, a, b, b, c, c are arranged in three rows and two columns, with each row and each column containing different letters, then there are ( ) possible arrangements.

A. 12 B. 17 C. 24 D. 36

B. Fill in the blanks

4 A railway station has 3 entrances and 5 exits. There are ________ possible choices for someone to enter and leave the station.

5 There are 21 boys and 24 girls in a class. If one boy and one girl are selected to participate in a school event, then there are ________ possible selections.

6 There are 3 routes from area A to area B and 4 routes from area B to area C. So there are ________ routes that can be taken from area A to area C via area B.

7 In a fashion show, there are 7 designs of hats, 12 designs of coats and 8 designs of trousers. If a model wears one hat, one coat and one pair of trousers each time, then there are ________ possible combinations of hats, coats and trousers that the model can wear.

8 A method of dividing a positive integer, n, into a sum of two or more positive integers is called partitioning, and the number of methods that can be used to split a number, n, is denoted by $p(n)$. For example, $3 = 1 + 2 = 1 + 1 + 1$, and $1 + 2$ is considered to be the same partitioning as $2 + 1$, so $p(3) = 2$. Therefore $p(6) =$ ________.

9 In the set of integers from 1 to 200 inclusive, there are ________ numbers that do not contain the digit 8.

C. Questions that require solutions

10 In the quadrilateral $ABCD$, there are 4, 3, 5 and 1 point marked on its four sides AB, BC, CD and DA respectively (excluding its four vertices). Take one point marked on each side as a vertex of a new quadrilateral, then how many different new quadrilaterals can it possibly obtain?

11 How many positive factors does 720 have?

12 If $f(x)$ is an nth degree polynomial and $g(x)$ is an mth degree polynomial, then what is the maximum number of terms of the product of $f(x)$ and $g(x)$? After collecting the like terms, what is the smallest number of terms?

13 A palindromic number refers to a positive integer that reads the same backwards as forwards, for example, 22, 121, 3443, 94 249, and so on. There are 9 two-digit palindromic numbers: 11, 22, 33, ⋯, 99, and there are 90 three-digit palindrome numbers: 101, 111, 121, ⋯, 191, 202, ⋯, 999.

(a) How many four-digit palindromic numbers are there?

(b) How many $2n + 1$-digit palindromic numbers are there? (n is a positive number)

7.2 Probability of mutually exclusive events

Learning objective

Use the property that the probabilities of an exhaustive set of mutually exclusive events sum to one.

A. Multiple choice questions

1. If events A and B are mutually exclusive, then (　　).

   A. $A \cup B$ is a certain event.

   B. $\overline{A} \cap \overline{B}$ is a certain event.

   C. $\overline{A}$ and $\overline{B}$ must not be mutually exclusive.

   D. $\overline{A}$ and $\overline{B}$ may or may not be mutually exclusive.

2. If four cards of different colours, red, yellow, black and white, are randomly given to four people, A, B, C and D, so that they have one each, then the event "A gets a white card" and the event "B gets a white card" are (　　).

   A. certain events

   B. complementary events

   C. mutually exclusive but not complementary events

   D. impossible events

3. A bag contains two red balls and two black balls. If two balls are taken out of the bag at random, then the two events that are mutually exclusive but not complementary are (　　).

   A. "there is at least one black ball" and "they are all black balls"

   B. "there is at least one black ball" and "there is at least one red ball"

   C. "it happens to be one black ball" and "it happens to be two black balls"

   D. "there is at least one black ball" and "they are all red balls"

4. In archery practice, if someone fires two arrows, one after the other, then the event (　　) and the event "hitting the target at least once" are mutually exclusive.

   A. "hitting the target at least once"

   B. "missing the target with both arrows"

C. "hitting the target with either arrow"
D. "hitting the target with one arrow only"

B. Fill in the blanks

5 When two players, A and B, play chess, the probability of a draw is $\frac{1}{2}$, while the probability of B winning is $\frac{1}{3}$. So the probability of A winning is ________.

6 In archery practice, if three arrows are fired, one after the other, then the complementary event of "hitting the target a maximum of two times" is ____________________.

7 A product is classified into one of three grades, A, B and C. A product graded B or C is regarded as defective. The probability of a grade B product being made is 0.03 and the probability of a grade C product being made is 0.01. A product is selected at random. What is the probability that this product is not defective? ________

8 The annual precipitation in an area has never been less than 100 mm. The table shows the probability that the annual precipitation in this area is within the following intervals.

Annual precipitation, p(mm)	$100 \leqslant p < 150$	$150 \leqslant p < 200$	$200 \leqslant p < 250$	$250 \leqslant p < 300$
Probability	0.21	0.16	0.13	0.12

The probability that the annual precipitation is greater than 300 mm is ________.

C. Questions that require solutions

9 In a mathematics test, the probability that Chen scores more than 90 marks out of 100 is 0.18, the probability that he scores 80–89 marks is 0.51, the probability that he scores 70–79 marks is 0.15, and the probability that he scores 60–69 marks is 0.09. Find the probability that Chen scores 80 marks or more in his mathematics test. If the pass mark is 60, find the probability that Chen passes the test.

10 In a competition, a competitor hits rings that give scores of 10, 9, 8, 7 and less than 7. The probabilities of achieving each score are 0.24, 0.28, 0.19, 0.16 and 0.13, respectively. Find:

(a) the probability of hitting a ring with a score of 10 or 9

(b) the probability of hitting a ring with a score of at least 7

(c) the probability of hitting a ring with a score of less than 8.

11 In a promotion in a shopping centre, 1 lottery ticket is given to a customer for every £100 they spend. Every 1000 lottery tickets contain one grand prize, 10 first prizes and 50 second prizes. The probabilities of one lottery ticket winning the grand prize, the first prize and the second prize are A, B and C, respectively.

(a) Find (i) $P(A)$ (ii) $P(B)$ (iii) $P(C)$.

(b) Find the probability of a ticket winning a prize.

(c) Find the probability of a ticket not winning either a grand prize or a first prize.

12 Bag A contains 3 white balls, 7 red balls and 15 black balls. Bag B contains 10 white balls, 6 red balls and 9 black balls. One ball is picked out of each bag at random. Find the probability that the two balls are the same colour.

7.3 Probability of independent and dependent combined events

Learning objective

Calculate the probability of independent and dependent combined events.

A. Multiple choice questions

1 When firing an arrow at a target, competitor A hits the target 8 times out of 10 and competitor B hits the target 7 times out of 10. If both A and B fire an arrow at the target, then the probability that they both hit the target is (　　).

A. $\frac{14}{25}$　　B. $\frac{12}{25}$　　C. $\frac{3}{4}$　　D. $\frac{3}{5}$

2 If events A and B are independent events, and $\overline{A}$ and $\overline{B}$ are the complementary events of A and B, respectively, then which of the following equations may not hold true? (　　)

A. $P(AB) = P(A)P(B)$　　B. $P(\overline{A}B) = P(\overline{A})P(B)$

C. $P(A + B) = P(A) + P(B)$　　D. $P(\overline{A} \cdot \overline{B}) = [1 - P(A)][1 - P(B)]$

3 A coin is flipped four times in a row. The probability of getting tails at least once is (　　).

A. $\frac{1}{16}$　　B. $\frac{3}{4}$　　C. $\frac{15}{16}$　　D. 1

4 The probability that a weather station accurately forecasts the weather is 0.8. The weather station predicts the weather for two days in a row. The probability of one of these forecasts being accurate is (　　).

A. 0.16　　B. 0.32　　C. 0.64　　D. 0.96

B. Fill in the blanks

5 The process of selecting a candidate for a position as chairperson of a club consists of two steps, A and B. If the pass rates of steps A and B are 95.7%, and 98.1%, respectively, then the pass rate of being selected is ________ (correct to 2 decimal places).

6 The probability of each of three athletes breaking a national record is 0.01. So the probability that all of them fail to break the national record in one Olympic Games is ________ (to 2 s. f.).

7 The probability of an archer hitting the target each time they fire an arrow is 0.9, and each result is independent. If the archer fires 4 arrows in a row and misses the target with the first arrow, then the probability of all the other 3 arrows hitting the target is ________.

8 The probabilities of two basketball players, A and B, scoring a point from a penalty shot are 0.7 and 0.6, respectively. If both A and B take a penalty shot, the probability of exactly one of these players scoring a point is ________.

C. Questions that require solutions

9 The probabilities of two batches of seeds, M and N, germinating are 0.8 and 0.7, respectively. One seed from each of the two batches is selected at random. Find:

(a) the probability that both seeds germinate

(b) the probability that at least one seed germinates

(c) the probability that exactly one seed germinates.

10 Three people are asked to decipher a password independently. The probabilities that each of them can decipher the password are $\frac{1}{5}$, $\frac{1}{4}$ and $\frac{1}{3}$, respectively. What is the probability of the password being deciphered?

11 A fair dice is thrown three times in a row. Use a tree diagram to find the probability that the first number obtained is even, the second is odd and the third is greater than 4.

12 An experiment is carried out four times. In each of these experiments, the probability of event A occurring is the same. If the probability that event A occurs at least once is $\frac{65}{81}$, find the probability of event A occurring in each experiment.

13 As shown in the diagram, three components, A, B, and C, are connected to form two systems, N_1 and N_2. When components A, B, and C work normally, system N_1 works normally; when component A works normally and at least one of components B or C works normally, system N_2 works normally. Given that the probabilities of components A, B, and C working normally are 0.80, 0.90, and 0.90, respectively, find the probabilities, P_1 and P_2, that systems N_1 and N_2 respectively, work normally.

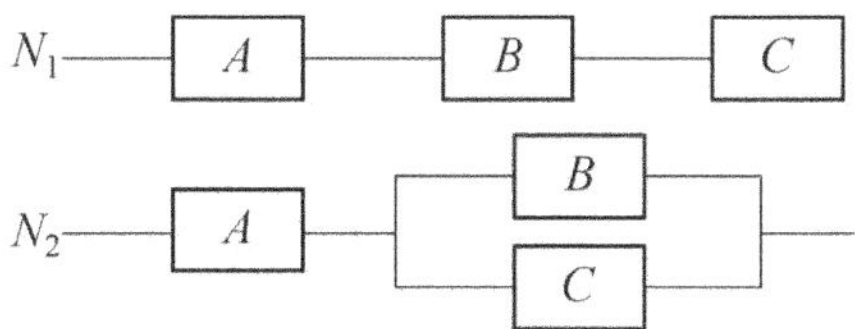

7.4 Theoretical probability and experimental probability

Learning objective

Construct theoretical possibility spaces for single and combined experiments with equally likely outcomes and use these to calculate theoretical probabilities.

A. Multiple choice questions

1 The frequency $\frac{m}{n}$ of random event A satisfies ( ).

A. $\frac{m}{n}=0$ B. $\frac{m}{n}=1$ C. $0<\frac{m}{n}<1$ D. $0\leqslant\frac{m}{n}\leqslant 1$

2 If a fair coin is flipped twice, the probability of getting heads both times is ( ).

A. $\frac{1}{4}$ B. $\frac{1}{2}$ C. $\frac{3}{4}$ D. 1

3 A fair dice is rolled three times and an even number is obtained each time. If the dice is rolled one more time, the probability of getting an even number is ( ).

A. 1 B. 0 C. $\frac{1}{2}$ D. uncertain

4 In a prize draw, 1000 lottery tickets are issued. 20 of these tickets are for first prize, 80 tickets are for second prize and 200 tickets are for third prize. The probability of the first lottery ticket issued winning a prize in the lottery is ( ).

A. $\frac{1}{50}$ B. $\frac{2}{25}$ C. $\frac{1}{5}$ D. $\frac{3}{10}$

B. Fill in the blanks

5 A card is selected at random from a pack of 52 cards. The probability that the card selected is a heart is ________.

6 A drawing pin is thrown 300 times and lands "point up" 216 times. So the frequency of "point up" is ________.

7 The table shows the frequency of the highest annual water level of a river over a certain period of time.

Range of highest annual water level (m)	$0 \leqslant m < 10$	$10 \leqslant m < 12$	$12 \leqslant m < 14$	$14 \leqslant m < 16$	$m \geqslant 16$
Frequency	0.10	0.28	0.38	0.16	0.08

If a "safe water level" is a height of 14 m or less, the relative frequency of "safe water level" is ________.

8 The number of newborn babies in a city over four years and the number of these that were boys are shown in the table.

(a) Complete the table to show the relative frequency of newborn baby boys.

	Year 1	**Year 2**	**Year 3**	**Year 4**
Number of newborn babies	5544	9013	13 520	17 191
Number of newborn baby boys	2716	4899	6812	8590
Frequency of newborn baby boys				

(b) According to the data shown in the table, the probability that a baby born in this city is a boy is about ________.

9 A box contains 10 cards numbered 1 to 10. A card is selected at random, its number is recorded, and it is then put back in the box. This is carried out 100 times. The table shows the results.

Number on card	1	2	3	4	5	6	7	8	9	10
Number of times picked	13	8	5	7	6	13	18	10	11	9

The frequency of randomly selecting an odd-numbered card is ________.

C. Questions that require solutions

10 A supermarket randomly selects 1000 customers and records their purchases of four products, A, B, C and D. The data are presented in the following statistical table, in which, "✓" means purchased and "×" means not purchased.

Number of customers \ Product	A	B	C	D
100	✓	×	✓	✓
217	×	✓	×	✓
200	✓	✓	✓	×
300	✓	×	✓	×
85	✓	×	×	×
98	×	✓	×	×

(a) Estimate the probability that a customer purchases both products B and C.

(b) Estimate the probability that a customer purchases three products at the same time.

(c) If a customer has purchased product A, then which of the products, B, C, and D, is the customer most likely to purchase?

11 A fish farm recorded the number of freshwater fish that hatched under the same conditions out of a batch of 10 000 eggs. They found that 8520 fish hatched.

(a) Find the relative frequency of a fish from this batch hatching.

(b) Estimate the number of freshwater fish that will hatch from 30 000 fish eggs.

(c) Estimate the number of fish eggs that are needed to hatch 5000 freshwater fish.

12 Table tennis balls produced by a company are used for an international table tennis game. Some samples of these balls are tested. The results are shown in the table.

Size of sample n	50	100	200	500	1000	2000
Number of good-quality balls m	45	92	194	470	954	1902
Relative frequency of good-quality balls $\frac{m}{n}$						

(a) Calculate the relative frequencies of good-quality balls from each batch and complete the table.

(b) One ball is randomly selected from each batch. Estimate the probability of getting a good-quality ball. Give your answer correct to 3 decimal places.

7.5 Conditional probabilities

Learning objective

Calculate and interpret conditional probabilities in different contexts.

A. Multiple choice questions

1 A box contains 20 identical balls, of which 5 are red, 5 are yellow and 10 are green. A ball is taken from the box at random. If the ball selected is not red, then the probability that it is green is ( ).

A. $\frac{5}{6}$ B. $\frac{3}{4}$ C. $\frac{2}{3}$ D. $\frac{1}{3}$

2 According to meteorological statistics over the years, in a region in April, the probability of an easterly wind is $\frac{9}{30}$, the probability of rain is $\frac{11}{30}$, and the probability of both rain and an easterly wind is $\frac{8}{30}$. Given that there is an easterly wind, the probability of rain is ( ).

A. $\frac{9}{11}$ B. $\frac{8}{11}$ C. $\frac{2}{5}$ D. $\frac{8}{9}$

3 10 identical cards are numbered 1 to 10. A card is randomly chosen, its number is recorded and another card is randomly chosen without replacement. The probability of obtaining two numbers whose product is an odd number is ( ).

A. $\frac{1}{2}$ B. $\frac{1}{3}$ C. $\frac{2}{3}$ D. $\frac{2}{9}$

B. Fill in the blanks

4 In a batch of products from a factory, the percentages of first class, second class and third class products in terms of quality are 60%, 35% and 5% respectively. Given that one of the products is selected at random and it is not in the third-class category, the probability of it being a first-class product is ________.

5 A company has 18 employees. 9 have blood type O, 3 have blood type A, 4 have blood type B and 2 have blood type AB. Two of these employees are selected at random. If one of these employees has blood type A, then the probability that the second employee has blood type O is ________.

6 Two people, A and B, each select at random a card with a number from 1 to 15 without replacement. Given that the number that person A takes is a multiple of 5, the probability that the number that person A takes is greater than the number that person B takes is ________.

7 A box contains 20 identical balls. 10 of these balls are red and 10 are white. A person selects two balls from the box at random without replacement. If the first ball selected is red, the probability that the second ball is white is ________.

8 There are two children in a family. It is equally likely that a child is a boy or a girl. Given that there is a girl in this family, then the probability that the other child is a boy is ________.

C. Questions that require solutions

9 The table shows the result of a class survey on the number of siblings each student has.

	No siblings	**1 sibling**	**2 siblings**
Female	7	6	2
Male	9	3	1

Find:

(a) the probability that a student in the class has 2 siblings

(b) the probability that a female student in the class has 2 siblings

(c) the probability that a male student in the class has no siblings.

10 Meteorological records over the last hundred years show the weather in two cities, M and N, in the same region. The proportions of the days when it was raining in M and N are 20% and 18%, respectively, and the proportion of the days when it rained simultaneously in both places is 12%.

(a) Given that it rained in city N, what is the probability that it also rained in city M?

(b) Given that it rained in city M, what is the probability that it also rained in city N?

11 Among 100 items produced in a factory, 70 items are classified as grade I and 25 items are classified as grade II. Both grade I and grade II are acceptable in terms of quality. One item produced by the factory is selected at random.

(a) What is the probability of the item being of grade I?

(b) Given that the item selected is acceptable, find the probability that it is of grade I.

12 Two fair dice are thrown together.

(a) When the sum of the numbers obtained is 7, what is the probability of one of the dice showing 2?

(b) When the two dice show different numbers, what is the probability that one of the dice shows 4?

Unit test 7

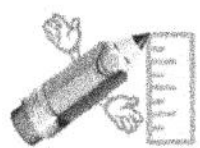

A. Multiple choice questions

1 Of the following statements, the correct one is (   ).

A. Of 500 people, there are at least two people who have the same birthday.
B. Of 500 people, there are at most two people who have the same birthday.
C. Of 500 people, it is impossible for any two people to have the same birthday.
D. Of 500 people, it is possible for two people to have the same birthday.

2 If a two-digit number is randomly selected, the probability of it being an integer that is a multiple of 10 is (   ).

A. $\frac{2}{9}$ B. $\frac{1}{9}$ C. $\frac{1}{10}$ D. $\frac{1}{5}$

3 If one card is randomly selected from a pack of 52 cards, the probability of selecting a king is (   ).

A. $\frac{1}{4}$ B. $\frac{1}{2}$ C. $\frac{1}{13}$ D. $\frac{1}{52}$

4 There are 5 screws in a box and 2 of them are damaged. If two of these screws are randomly selected from the box, then $\frac{7}{10}$ is equal to (   ).

A. the probability that one screw is damaged
B. the probability that both screws are damaged
C. the probability that both screws are in good condition
D. the probability that at least one screw is damaged

B. Fill in the blanks

5 Three people, A, B and C, are standing in a row. The probability that person A is standing in the middle is ________.

6 A bag contains 4 white balls, 1 red ball and 7 yellow balls. The probability that a randomly selected ball is white is ________.

7 A small cube has six faces marked with a parallelogram, a circle, an isosceles trapezium, a rhombus, an equilateral triangle and a right-angled trapezium. The cube is thrown once. The probability that the cube lands showing a shape with both rotational symmetry and line symmetry is ________.

8 If two different numbers are chosen from the digits 1, 2, 3, 4 and 5 to form a two-digit number, then the probability that the two-digit number is greater than 40 is ________.

9 One person is chosen at random from a group of people. If the probability that the person chosen is less than 160 cm tall is 0.2 and the probability that the person chosen is between 160 cm and 175 cm tall is 0.5, then the probability that the person chosen is taller than 175 cm is ________.

10 A bag contains 9 white balls and 2 red balls. 3 balls are selected randomly from the bag. Consider the following pairs of events.

① "There is 1 red ball" and "All are white balls"

② "There is at least 1 red ball" and "All are white balls"

③ "There is at least 1 red ball" and "There are at least 2 white balls"

④ "There is at least 1 red ball" and "There is at least 1 white ball"

The events described in ________ are complementary events.

11 Look at the following three propositions.

① A factory produces a large batch of a product. The probability of an item from the factory being defective is 10%. If 100 pieces of the product are randomly selected, then 10 of them must be defective.

② A coin is flipped 7 times and it lands on heads 3 times. Therefore, the probability of getting heads on this coin is $\frac{3}{7}$.

③ The frequency at which a random event occurs is the probability of the occurrence of this random event.

The proposition(s) described in ________ is/are false.

12 The probability that an animal of a certain species lives up to 20 years old is 0.8, and the probability that it lives up to 25 years old is 0.4. Given that an animal of the species is now 20 years old, the probability of it living up to 25 years old is ________.

13 The probability of a basketball player scoring a field goal is 40% each time they take a shot. A random simulation method is used to estimate the probability that the player will score two goals from three shots. Random integers from 0 to 9 are generated using a calculator. The number of goals scored is represented by 1, 2, 3 and 4 and the number of times the player misses is represented by 5, 6, 7, 8, 9 and 0. Every 3 randomly generated numbers are taken as a group to represent the results for the three shots. The following 20 sets of randomly generated numbers are obtained by this simulation.

907 966 191 925 271 932 812 458 569 683
431 257 393 027 556 488 730 113 537 989

Based on this estimate, the probability of the basketball player getting two goals from three shots is ________.

C. Questions that require solutions

14 Two fair dice with six faces marked 1 to 6 are rolled.

(a) What is the probability that the product of the two numbers rolled is larger than 25?

(b) When the product of the two numbers rolled is 12, what is the probability of one of the dice showing the number 3?

(c) When the two dice show different numbers, what is the probability of one of the dice showing an even number?

15 15% of the students in a class failed a mathematics test and 5% failed an English test. 3% failed in both subjects.

(a) If a student failed the mathematics test, find the probability that they also failed the English test.

(b) If a student failed the English test, find the probability that they also failed the mathematics test.

16 Ashwin analysed the weather in a city in April of a randomly selected year. The results are shown in the table.

Date	1	2	3	4	5	6	7	8	9	10
Weather	sunny	rainy	cloudy	cloudy	cloudy	rainy	cloudy	sunny	sunny	sunny
Date	11	12	13	14	15	16	17	18	19	20
Weather	cloudy	sunny	sunny	sunny	sunny	sunny	cloudy	rainy	cloudy	cloudy
Date	21	22	23	24	25	26	27	28	29	30
Weather	sunny	cloudy	sunny	sunny	sunny	cloudy	sunny	sunny	sunny	rainy

(a) Using the data given in the table, estimate the probability that it will not rain on a day in April in this city.

(b) A school in the city plans to have a two-day sports event starting on a sunny day in April. If these two days are randomly chosen, estimate the probability that there will be no rain during the games.

17 To prepare for an international competition, a national archery team have gone through an intensive period of training. After the training, the probability of a team member scoring 7–10 points is shown in the table below. Scores below 7 have not been recorded.

Number of points scored	10	9	8	7
Probability	0.32	0.28	0.18	0.12

Each team member takes one shot.

Find the probability that a randomly selected team member:

(a) scores 9 or 10 points

(b) scores less than 8 points.

Chapter 8 Statistics

8.1 Sample and population

Learning objective

Understand and use sampling to describe populations.

A. Multiple choice questions

1 Which of the following statements describes the relationship between sample and population incorrectly? (　　)

A. A sample is a part of the population.

B. A sample is a subset of the population.

C. A sample is selected from the population.

D. A sample has nothing to do with the population.

2 There are 215 passengers on an aeroplane from Shanghai to London. The airline wants to find out how satisfied the passengers are with their service, and hence decide to interview 40 of the passengers. Which of the following methods of selection will be most likely to produce a random sample? (　　)

A. Selecting the first 40 passengers who board the aeroplane.

B. Selecting any 40 passengers who are in the "Frequent Flying Program".

C. Selecting any 40 passengers who are between 40 to 45 years old.

D. Writing the names of all the passengers on slips of paper which are identical, putting the slips into a box, mixing them well, and then picking 40 of them.

3 Lee is an organiser of a healthy eating club. He wishes to know the popularity of a vegetarian diet within a community during January. He randomly selects a restaurant every evening in the month, and then interviews the 10th guest entering the restaurant. Which of the following methods is he using to select the sample? (　　)

A. stratified sampling　　B. cluster sampling

C. random sampling　　D. systematic sampling

4 A regional education authority wishes to gather information about students' scores in a college entrance examination. From the 248 high schools in the region, three schools are randomly selected and the data of all the students from these schools are used in the sample. Which of the following methods is being used to select the sample? ( )

A. stratified sampling B. cluster sampling

C. random sampling D. systematic sampling

5 Which of the following methods allows us to identify a cause-and-effect relationship between two variables? ( )

A. a sample survey B. a census

C. a well-designed experiment D. an observational study

B. Fill in the blanks

6 The purpose of sampling is to make inferences about the ________.

7 The most commonly used sampling technique used for making inferences about a population is ________. (Choose from: stratified sampling cluster sampling random sampling systematic sampling.)

8 A survey is carried out with 50 students selected from each year group from Year 7 to Year 11 of a secondary school. Identify the population and the sample. The population is ________________ in the school, and the sample is ________________.

C. Questions that require solutions

9 Describe the population and the sample in each of the following scenarios.

(a) In a survey, 868 adults across a country were asked if they thought there was evidence of the existence of UFOs. The results revealed that 634 of the adults said yes.

(b) A survey of 1220 families in a city found that 59% of them think that buying a home is the best investment that a family can make.

10 In a secondary school, the head of the maths department wishes to select a stratified sample, consisting of 15% of all the students, for a survey. The table shows the number of students in each year group in the school. How many students should be selected from each year group for the survey? Show your working.

Year group	7	8	9	10	11
Number of students	242	236	253	249	240

11 Explain briefly why selecting a sample, instead of the whole population, is often necessary.

8.2 Tables and line graphs for time series data

Learning objective

Interpret and construct tables and line graphs for time series data.

A. Multiple choice questions

1 The line graph below shows the daily maximum temperatures (℃) in a city for a period of 6 months. From the graph, the biggest variation between two consecutive months was from (　　).

A. January to February　　B. February to March

C. March to April　　D. April to May

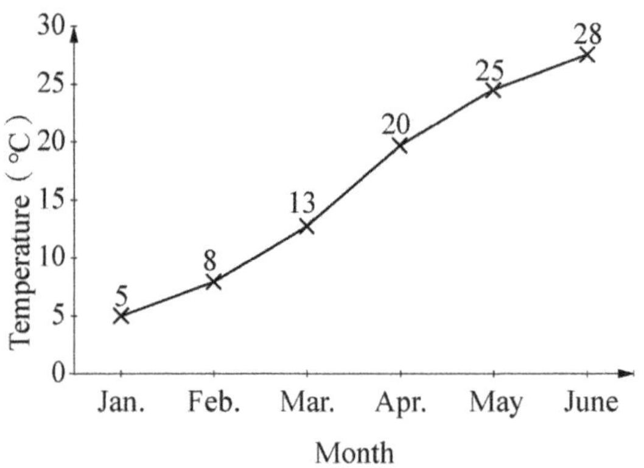

2 The table below shows the average fertility rate in England and Wales from 2011 to 2020 (Source: Office for National Statistics – Births in England and Wales).

Year	**2011**	**2012**	**2013**	**2014**	**2015**	**2016**	**2017**	**2018**	**2019**	**2020**
Fertility rate	1.93	1.94	1.85	1.83	1.82	1.81	1.76	1.70	1.66	1.58

What was the biggest difference in fertility rates between two consecutive years from 2011 to 2020? (　　)

A. 0.01　　B. 0.17　　C. 0.09　　D. 0.18

3 Use the data in the table given in Question 2. What was the mean fertility rate for the years 2012 to 2015? (   )

A. 1.86　　B. 1.81　　C. 1.92　　D. 1.89

4 Use the data in the table given in Question 2. The mean fertility rate for the ten years is (   ).

A. 1.82　　B. 1.81

C. 1.815　　D. none of the above

B. Fill in the blanks

5 Look at the following line graph, which represents the sales of ballpoint pens in a shop from 2014 to 2018.

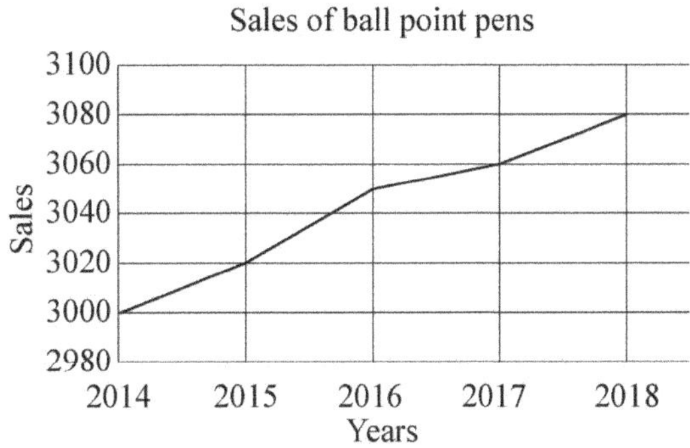

(a) From 2014 to 2018, the number of sales increased by ________% (correct to one decimal place).

(b) The biggest increase was from year ________ to year ________.

(c) The smallest increase was from year ________ to year ________.

6 The total value of goods (in £1000s) produced by a manufacturer each quarter in 2020 is given in the table below. Complete the table.

Year 2020	Value of goods (in £1000s)	Cumulative sales
1st quarter	236	
2nd quarter		528
3rd quarter	287	
4th quarter		1208

C. Questions that require solutions

7 Refer to the table given in Question 2.

(a) Construct a line graph using the data.

(b) (i) Which statistical tool is the most useful when calculating the mean fertility rate across the ten years?

(ii) Which statistical tool, the table or the line graph, makes it easier to tell whether there is a steady increase over the ten years? Give a reason for your answer.

8 The table below shows the Gross Domestic Product (GDP) of the UK from 2016 to 2021 (Source: UK Office for National Statistics).

Year	2016	2017	2018	2019	2020	2021
GDP (in billion pounds)	2137	2182	2218	2255	2046	2198
Growth rate						

(a) Complete the table by filling in the growth rate for each year to the nearest 0.01% (Note: the GDP in 2015 is £2089 billion).

(b) Construct a line graph for the data given.

(c) What is the growth rate of the GDP from year 2016 to 2021? What is the mean annual growth rate? (to the nearest 0.01%)

(d) (i) Between which two years does the GDP increase the most?

(ii) Between which two years, does the GDP decrease? Give one possible reason for the decrease.

9 The following table shows the US unemployment rate over a 12-year period from 2007 to 2018 (Source: US Bureau of Labor Statistics).

Year	**2007**	**2008**	**2009**	**2010**	**2011**	**2012**	**2013**	**2014**	**2015**	**2016**	**2017**	**2018**
Unemployment rate	5.0	7.3	9.9	9.3	8.5	7.9	6.7	5.6	5.0	4.7	4.1	3.9

(a) Construct a line graph to represent the data in the table.

(b) (i) Which year had the highest rate of unemployment? What was the value in that year?

(ii) Which year had the lowest rate of unemployment? What was the value in that year? (Give your answer to the nearest 0.1%.)

(c) During which period did the unemployment rate steadily decrease? State the year in which this decrease started. Give a reason for your answer.

8.3 Statistical graphs for grouped discrete data

Learning objective

Construct and interpret histograms and cumulative frequency graphs for grouped discrete data and continuous data.

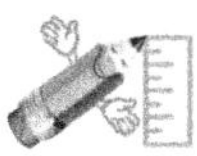

A. Multiple choice questions

1 200 students took a maths test in which the total mark was 100. The following cumulative frequency graph shows the marks the students scored.

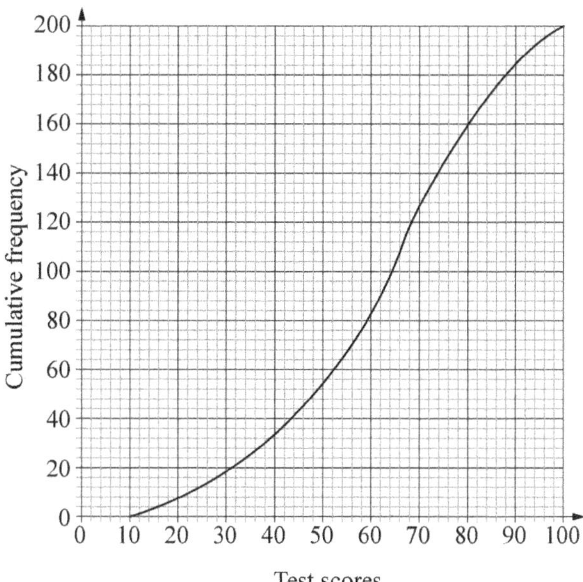

The median score was (   ) marks.

A. 50   B. 52   C. 64   D. 100

2 Refer to the cumulative frequency graph given in Question 1. The number of students who scored 70 marks or fewer in the test was (   ).

A. 43   B. 57   C. 74   D. 126

3 Refer to the cumulative frequency graph given in Question 1. Students who took the test were awarded grades, with grade A being the highest achievable. Given that 10% of the students obtained grade A, the minimum score required to obtain an A grade is (   ).

A. 32   B. 88   C. 90   D. 95

4 Which of the following types of statistical graph is most suitable for identifying the central tendency of a set of discrete data? (   )

A. histogram   B. pie chart   C. scatter diagram   D. bar chart

5 In which of the following histograms is the mean larger than the median? (   )

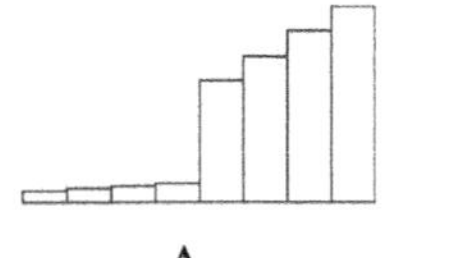

A.

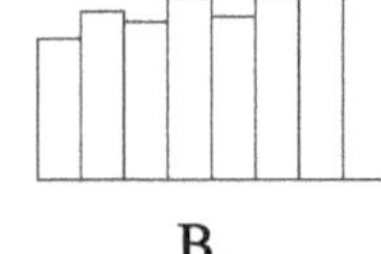

B.

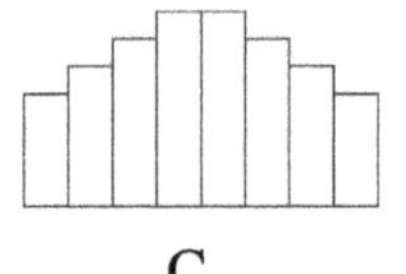

C.

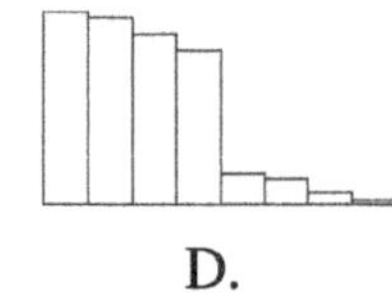

D.

B. Fill in the blanks

6 The following histogram with unequal class intervals shows the distribution of times that students spent completing a task.

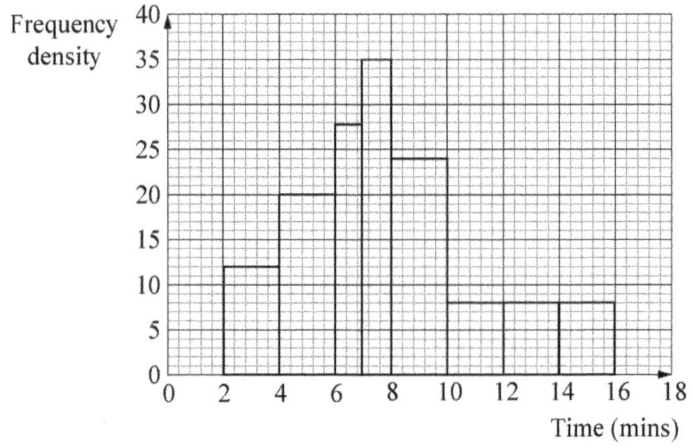

Complete the following frequency table using the data represented in the histogram.

Time (t minutes)	$2 < t \leqslant 4$	$4 < t \leqslant 6$	$6 < t \leqslant 7$	$7 < t \leqslant 8$	$8 < t \leqslant 10$	$10 < t \leqslant 16$
Frequency	24				48	

7 Refer to the histogram given in Question 6. There were ________ students who spent more than 7 minutes but no more than 8 minutes to complete the task.

8 Refer to the histogram given in Question 6. There were ________ students who spent no more than 6 minutes and ________ students who spent more than 8 minutes to complete the task, respectively.

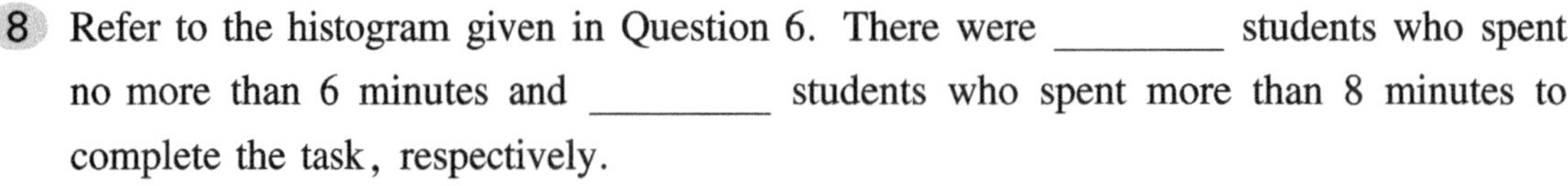

9 Refer to the histogram given in Question 6. There were ________ students in total who completed the task.

C. Questions that require solutions

10 A sample of cars is randomly selected from all the cars manufactured in a factory over a certain period of time. The length (in metres) of each car selected is measured. The results are summarised in the following table.

Length of car, x (metres)	**Frequency**	**Frequency density**
$2.80 \leqslant x < 3.00$	17	85
$3.00 \leqslant x < 3.10$	a	240
$3.10 \leqslant x < 3.20$	19	190
$3.20 \leqslant x < 3.40$	8	b

(a) Find the value of a and the value of b.

(b) Construct a histogram using the data.

(c) What conclusion can you draw by observing the histogram?

11 The arrival time of 204 students at an event was recorded. The number of minutes that each student was late was grouped and is shown in the table below.

Number of minutes late, t	$-2 \leqslant t < 0$	$0 \leqslant t < 2$	$2 \leqslant t < 4$	$4 \leqslant t < 6$	$6 \leqslant t < 10$
Number of students	43	51	69	22	19

(a) Explain what $-2 \leqslant t < 0$ means in this context.

(b) (i) Draw a cumulative frequency graph for the data in the table.

(ii) Use your graph from (b)(i) to estimate the median and the interquartile range of the number of minutes that students arrived late.

8.4 Graphical representation of the distributions of statistical data (1)

Learning objective

Interpret, analyse and compare data sets using appropriate measures of central tendency and spread and statistical diagrams, including box plots.

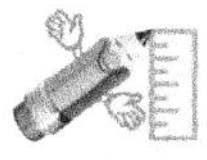

A. Multiple choice questions

1 A biologist collected 100 leaves from a plum tree and 100 leaves from a peach tree. He measured their lengths to the nearest cm. The results are shown in the following box plots.

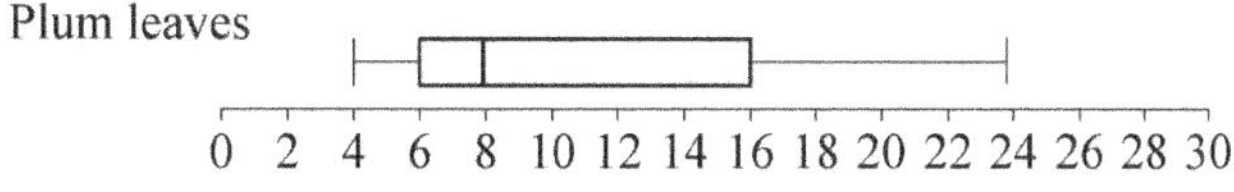

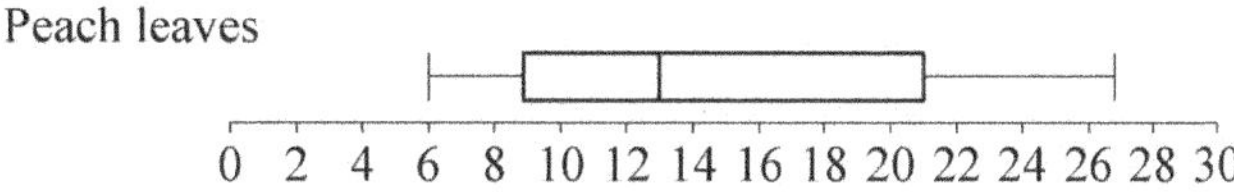

The range of the lengths of all 200 leaves is (　　).

A. 18　　B. 20　　C. 21　　D. 23

2 Refer to Question 1. Which the following cumulative frequency graphs is the best representation of the lengths of the plum leaves? (　　)

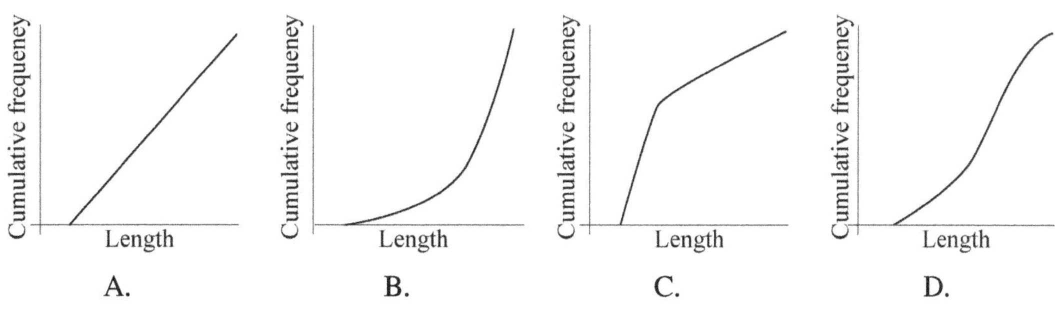

3 Refer to Question 1. Which of the following statements is incorrect? (　　)

A. The median of plum leaves is less than the median of peach leaves.

B. The range of plum leaves is less than the range of peach leaves.

C. The interquartile range of plum leaves is greater than that of peach leaves.

D. About 75% of peach leaves are greater than 50% of plum leaves.

4 Which of the following box plots represents data for which the mean is larger than the median? (   )

B. Fill in the blanks

5 The masses of 9 baseball players (in kilograms), from lightest to heaviest, are 69, 72, 75, 75, 78, 79, 80, 85 and 89. A box plot for this set of data is shown below. Give the masses represented by each letter on the box plot and find the interquartile range.

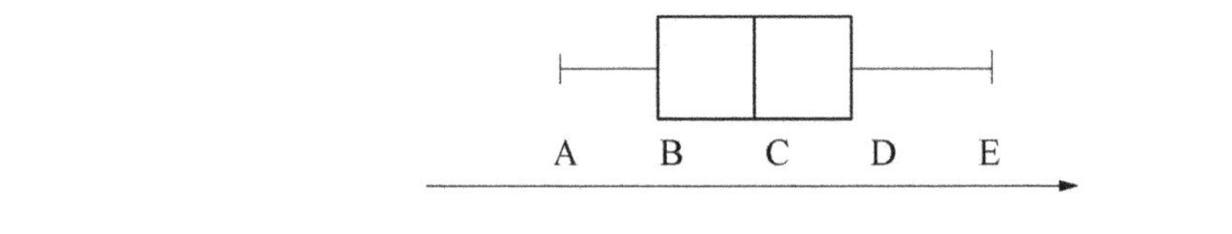

A = _______kg, B = _______kg, C = _______kg, D = _______kg, E = _______kg.
The inter-quartile range is _______.

6 Three populations A, B and C are of the same size and have the same range. Frequency diagrams for the three populations are given below. Match each box plot with the correct frequency diagram. (Write the correct population, A, B or C, on the line below each box plot.)

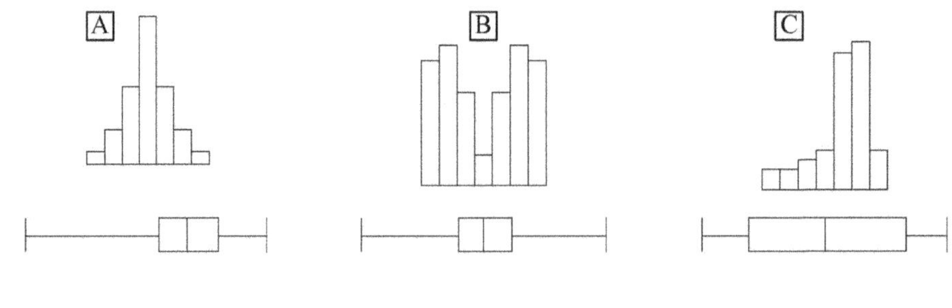

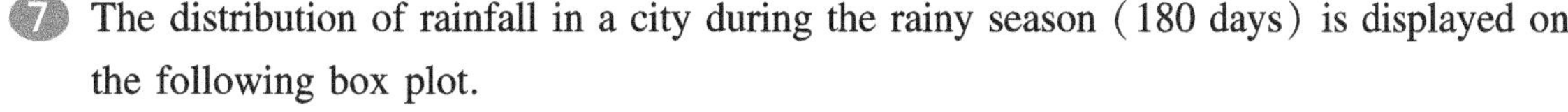

7 The distribution of rainfall in a city during the rainy season (180 days) is displayed on the following box plot.

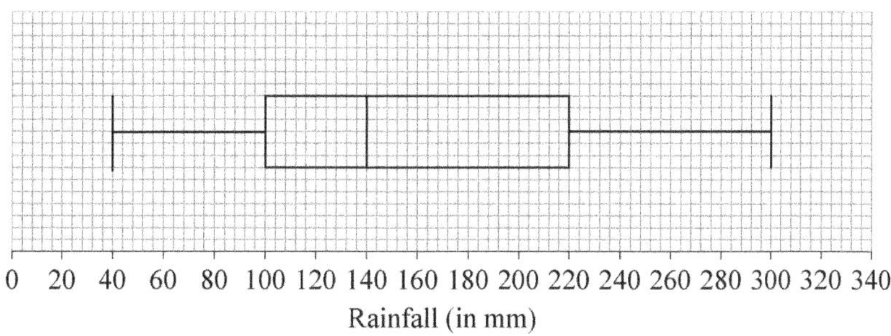

(a) Find the largest number of millimetres of rainfall.
(b) State the statistical term corresponding to the value of 140 mm.
(c) Write down the number of days on which the rainfall is between 100 mm and 220 mm.
(d) Find the percentage of days on which the rainfall is more than 140 mm.

8 In a class survey, students were asked about the amount of time each week that they spend watching videos on a social media platform. It was found that the median time was 3 hours 36 minutes, the upper quartile was 4 hours 42 minutes and the interquartile range was 3 hours 48 minutes. The most amount of time spent was 5 hours 12 minutes and the least amount of time was 30 minutes.

(a) Find the lower quartile.
(b) Represent this information on a box plot, using a scale of 2 cm to represent 60 minutes.

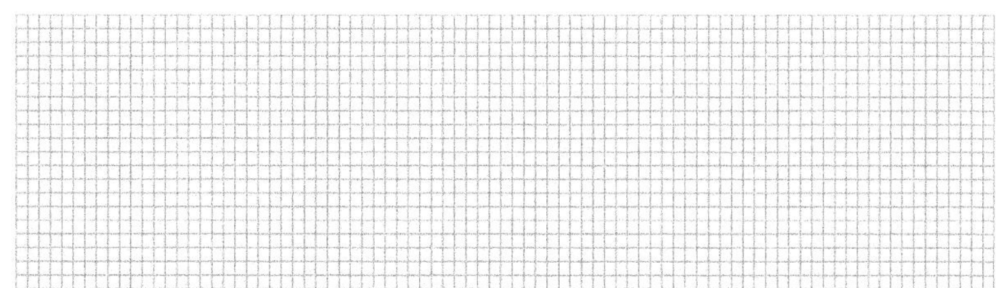

9. The following data represent the masses, in kilograms, of the swimmers in two swimming teams, team A and team B.

Team A: 65, 67, 69, 71, 63, 62, 70, 72, 66

Team B: 65, 67, 67, 68, 70, 64, 64, 65, 65

(a) Represent the data by drawing two parallel box plots.

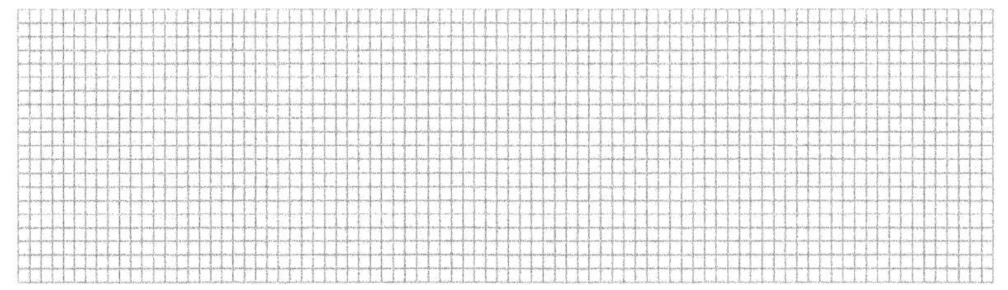

(b) Find the range, interquartile range and median for each set of data. Compare the masses of these two teams.

8.5 Graphical representation of the distributions of statistical data (2)

Learning objective

Interpret, analyse and compare data sets using appropriate measures of central tendency and spread and statistical diagrams and tables, including cumulative frequency graphs and frequency tables.

A. Multiple choice questions

1 Which of the following measures the central tendency of a set of data? ( )

A. mean and median
B. interquartile range
C. range
D. none of them

2 Which of the following measures the spread of a set of data? ( )

A. mean
B. median and mode
C. interquartile range
D. quartiles

3 10 employees in an IT company are randomly selected in a survey to find the typical level of salary. The results show that their monthly salaries (in £) are 3200, 3400, 25 400, 3200, 4100, 4300, 3600, 3800, 4200 and 3300. Which measure of central tendency is most appropriate to represent this set of data? ( )

A. mean
B. median
C. mode
D. none of the above

4 The heights of 200 pupils are collected in a survey. The summary statistics are as follows: mean: 116 cm, median: 115 cm, lower quartile: 108 cm, and upper quartile: 122 cm. Therefore, in the sample, there are about 100 pupils whose heights are ( ).

A. less than 116 cm
B. less than 108 cm
C. between 115 cm and 122 cm
D. between 108 cm and 122 cm

5 Let X_1, X_2, $\cdots$, X_n be n observations. If a constant a is added to each observation, then which of the following measures will not change? ( )

A. interquartile range
B. median
C. mean
D. lower quartile

B. Fill in the blanks

6 The modal class for the following frequency table is ________.

Mark	Frequency
$0 \leqslant x < 5$	4
$5 \leqslant x < 15$	7
$15 \leqslant x < 20$	16
$20 \leqslant x < 35$	6

7 A box contains 51 cards. Each card has a number from one to six written on it. The following frequency table shows how many cards there are with each number.

Number	1	2	3	4	5	6
Frequency	9	7	8	5	9	13

The median number on a card is ________ and the interquartile range is ________.

8 A data set is listed in ascending order: 3, 6, 7, 7, 8, x, 12, 13, y, 14, 15 and 16. The median of the data is 11 and the upper quartile is 13.5. Therefore $x =$ ________ and $y =$ ________.

C. Questions that require solutions

9 Consider four positive integers a, b, c and d, with $a < b < c \leqslant d$. The mode of these integers is 8, the median is 7 and the range is 5. Find the values of a, b, c and d.

10 The lower quartile Q_1 of a data set is 13 and the upper quartile Q_3 is 19. An outlier is defined as any data value which is more than 1.5 times the interquartile range above Q_3, or more than 1.5 times the interquartile range below Q_1. What is the smallest value below Q_1 in the data set that would be an outlier? What is the largest value above Q_3 in the data set that would be an outlier?

11 The following cumulative frequency graph shows the heights (in cm) of 200 young apple trees.

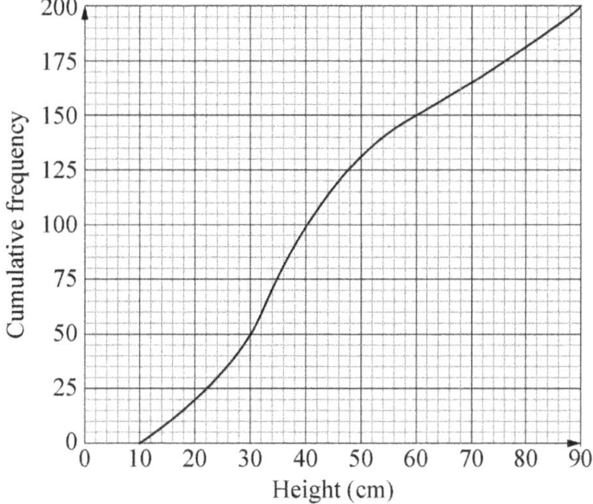

(a) Write down the median height of the trees.

(b) Write down the lower quartile and upper quartile of the heights.

(c) Estimate the number of trees that are more than 70 cm in height.

(d) The heights of the 200 trees are grouped as shown in the table below. Complete the table.

Height (cm)	**Frequency**
$10 \leqslant h < 30$	50
$30 \leqslant h < 50$	
$50 \leqslant h < 70$	35
$70 \leqslant h < 90$	

(e) Use the above table to estimate the mean height of the trees.

8.6 Apply statistics to describe a population

Learning objective

Select and apply statistical techniques to make inferences and draw conclusions when describing a population.

A. Multiple choice questions

1 Which of the following statements is/are correct? (　　)

① The range of a sample data set is never greater than the range of the population.

② The interquartile range is half the distance between the first quartile and the third quartile.

③ While the range is affected by outliers, the interquartile range is not.

A. ① only　　B. ③ only　　C. ① and ③ only　　D. ①, ② and ③

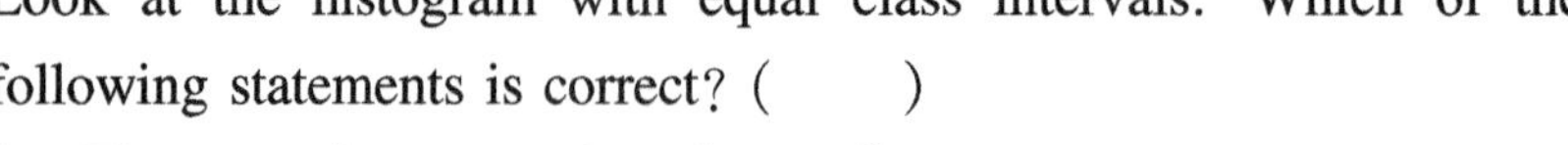

2 Look at the histogram with equal class intervals. Which of the following statements is correct? (　　)

A. The mean is greater than the median.

B. The mean is less than the median.

C. The mode is less than the mean.

D. The mode is greater than the mean.

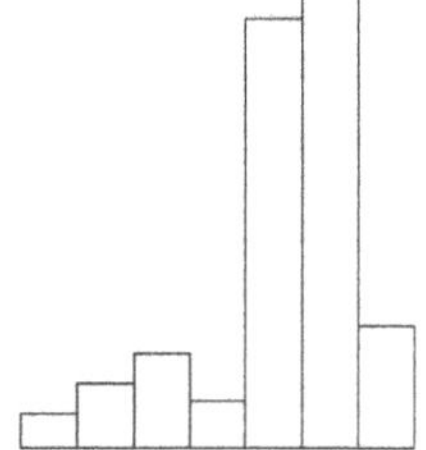

3 Which of the following statements about a population and a sample of the population is correct? (　　)

A. The mean of the population is always larger than the mean of the sample.

B. The mean of the population is always less than the mean of the sample.

C. The sample mean is always equal to the population mean.

D. None of the above is correct.

4 The five summary statistics of a data set are shown below. Which of the following statements is correct? (　　)

Minimum value	Lower quartile	Median	Upper quartile	Maximum value
2	5	8	12	25

A. The mean is larger than 8.
B. The mean is less than 8.
C. The mean is equal to 8.
D. None of the above is correct.

B. Fill in the blanks

5 When describing a statistical distribution, the three most commonly used key concepts are shape, ________ and ________.

6 The mean score on a test is 75. If each score is increased by 5, then the new mean is ________; if each score is increased by 5%, then the new mean is ________.

7 It is __________ that two different sets of data have the same five-number statistical summary. (Choose either: possible or impossible.)

8 It is _________ that two sets of data with the same box plots must have the same distribution. (Choose either: true or not true.)

C. Questions that require solutions

9 Fatima works in a shoe shop. The table below shows information about the shoe sizes of 100 customers who bought shoes from the shop last month.

Shoe size	**No. of customers**
6	15
7	39
8	26
9	20

(a) Find the median shoe size.

(b) Fatima needs to order 1500 pairs of shoes for a promotion next month. How many pairs of shoes with size 6 should she order? Justify your answer.

10 100 students took part in an essay writing competition. All the essays submitted were sent to two evaluators Robert and Nala. These two evaluators marked the essays independently. The following cumulative graph shows the results of their marking.

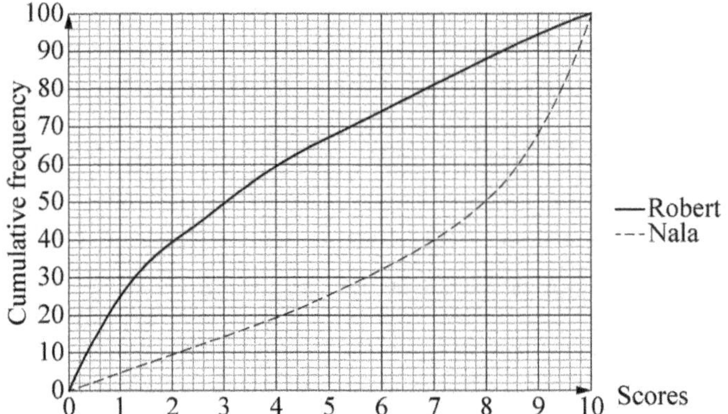

(a) Find the median and interquartile range of each of the two evaluators' scores.

(b) Use suitable data from these graphs to compare the central tendencies and spreads of the marks scored by the two evaluators.

(c) The students whose scores are greater than 9 are be awarded a medal, but there are only 10 medals available. Which evaluator's score would you use to decide who should get a medal? Explain your choice.

8.7 Scatter graphs of bivariate data (1)

Learning objective

Use, construct and interpret scatter diagrams of bivariate data, identifying correlation and drawing lines of best fit to describe trends in the data.

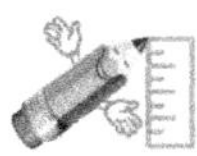

A. Multiple choice questions

1 Which of the following statistical graphs is the most commonly used to study the relationship between two variables? (　　)

A. histogram　　B. pie chart　　C. scatter diagram　　D. bar chart

2 A perfect positive correlation means (　　).

A. the points in a scatter diagram lie on an upward sloping line

B. the points in a scatter diagram lie on a downward sloping line

C. there is a direct cause-and-effect relationship between the variables

D. none of the above

3 Which of the following data sets shows a strong linear negative correlation? (　　)

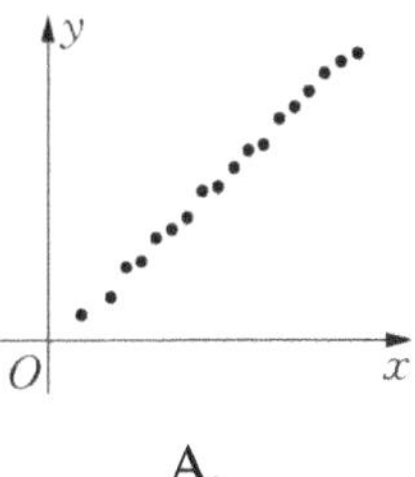

A.

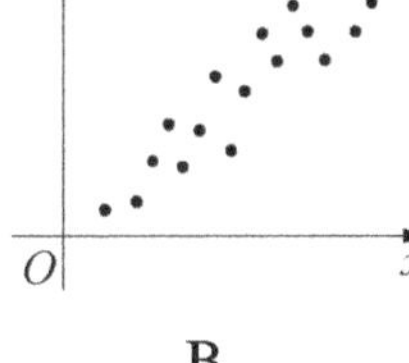

B.

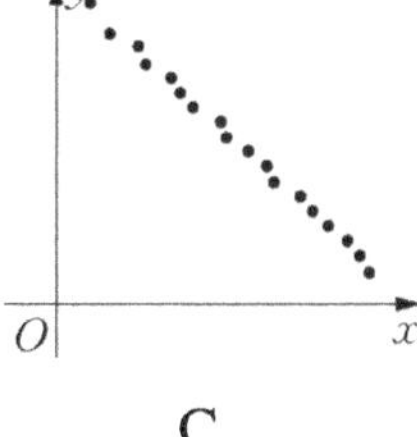

C.

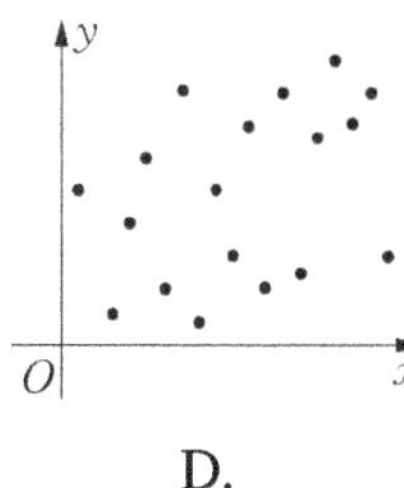

D.

4 A sociologist thinks that there is a correlation between income y (in £1000) per year and the number of years of education x that a person receives. They plotted their data on a scatter graph and obtained an equation for the line of best fit: $y = 11.8 + 7.42x$. Which of the following statements best describes the meaning of the slope of the line of best fit? (　　)

A. For each increase in education of 1 year, the estimated income per year increases by 11.8.

B. For each increase in education of 1 year, the estimated income per year increases by 7.42.

C. For each increase in income of £1000 per year, the estimated number of years of education increases by 11.8.

D. For each increase in income of £1000 per year, the estimated number of years of education increases by 7.42.

5 If every boy in a class has a brother who is exactly 3 cm taller than him, then there is (   ) correlation between the heights of boys and their brothers.

A. somewhat positive
B. somewhat negative
C. no correlation
D. perfectly positive

B. Fill in the blanks

6 To draw a scatter graph, you plot the ________ variable on the horizontal axis and the ________ variable on the vertical axis. (Choose from: independent or dependent.)

7 If the independent and dependent variables show a positive correlation, then as the independent variable increases, the dependent variable ________.

8 If the independent and dependent variables show a negative correlation, then as the independent variable increases, the dependent variable ________.

C. Questions that require solutions

9 The following table shows the heights and masses of ten students in a Year 11 class.

Height (cm)	182	173	162	190	178	168	180	175	185	183
Mass (kg)	73	68	60	79	66	64	80	65	75	73

(a) Construct a scatter diagram.

(b) What can you say about the direction of the association between height and mass of these students?

(c) What can you say about the relationship between the heights and masses of these students?

(d) What can you say about the strength of the relationship?

10 The table below shows the amount of fuel, y litres, remaining in the tank of a car and the distance, x kilometres, the car travelled after the tank was fully filled.

Distance travelled (km)	0	207	259	470	640	800
Amount of fuel (litres)	55	43	30	24	10	6

The graph below shows the first three results.

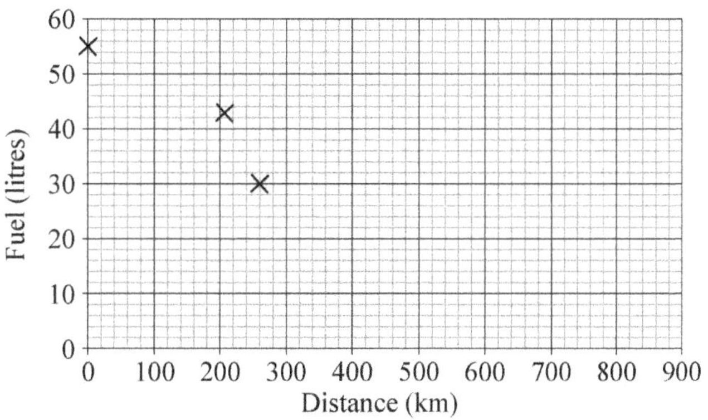

(a) Plot the last three results on the graph.
(b) Find the mean distance travelled, and the mean amount of fuel in the tank.
(c) Plot the mean point on the graph and sketch a line of best fit through the mean point.
(d) A car has travelled 320 km. Use the line of best fit to estimate the amount of fuel remaining in the tank.

8.8 Scatter graphs of bivariate data (2)

Learning objective

Identify correlation on a scatter diagram and know that it does not indicate causation; Use linear regression to interpolate and extrapolate trends.

A. Multiple choice questions

1 In a line of best fit, which is also called a linear regression line, $y = a + bx$, the slope b represents ( ).

A. the value of y where the regression line intersects the y-axis

B. the value of x where the regression line intersects the x-axis

C. the change in the dependent variable due to a one-unit change in the independent variable

D. the change in the independent variable due to a one-unit change in the response variable

2 The scatter graph showing the temperature (in ℃) plotted against the sales of ice creams (in £) in a shop is shown below. Which of the following can be said about the relationship between the temperature and the sales of ice creams? ( )

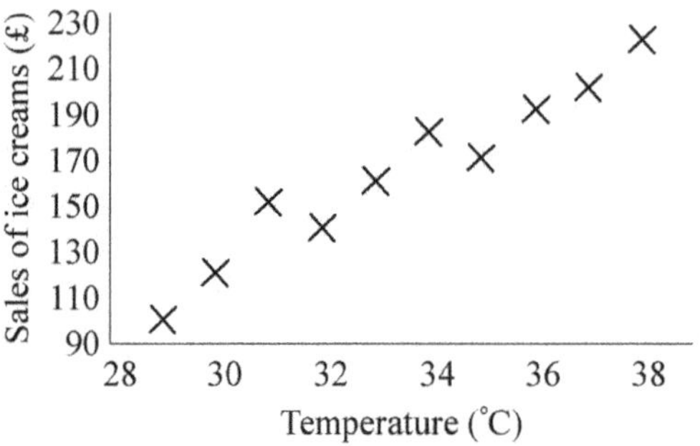

A. Higher temperatures are related to lower sales of ice creams.

B. Lower temperatures are related to higher sales of ice creams.

C. Higher temperatures are related to higher sales of ice creams.

D. Higher sales of ice creams are caused by higher temperatures.

3 If the correlation between two variables is negative, then which of the following statements is possible for any two points on a scatter graph? ( )

A. The first point has a larger x-value and a smaller y-value than the second point.
B. The first point has a larger x-value and a larger y-value than the second point.
C. The first point has a smaller x-value and a larger y-value than the second point.
D. All the above are possible.

B. Fill in the blanks

For each questions, 4–7, look at the scatter diagram given, and complete the statement with one of the following:

a strong positive correlation   a weak positive correlation

no correlation   a weak negative correlation   a strong negative correlation

4 There is ______________ between the two variables shown in the scatter diagram.

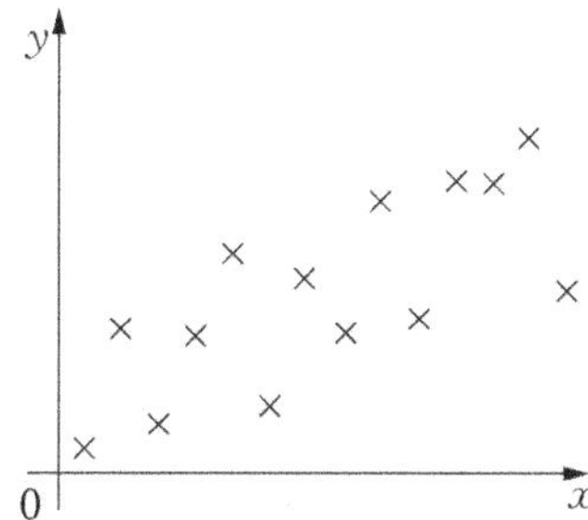

5 There is ______________ between the two variables shown in the scatter diagram.

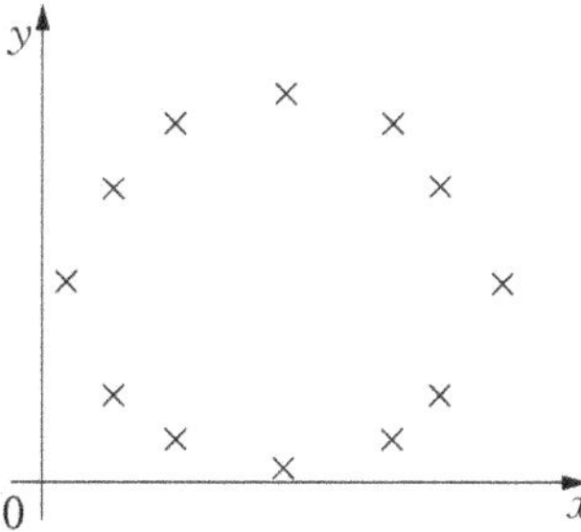

6 There is ______________ between the two variables shown in the scatter diagram.

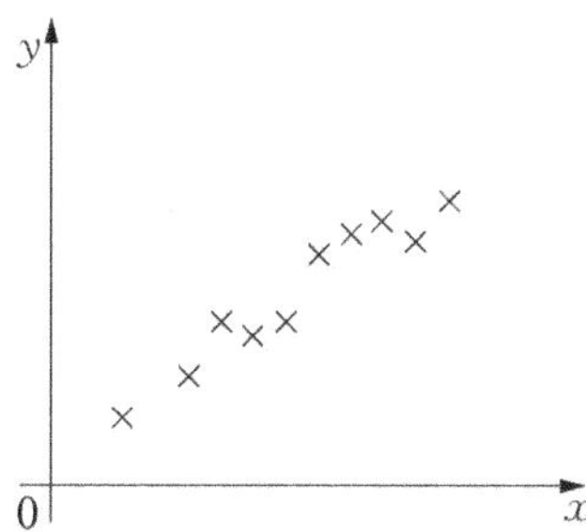

7 There is ________________ between the two variables shown in the scatter diagram.

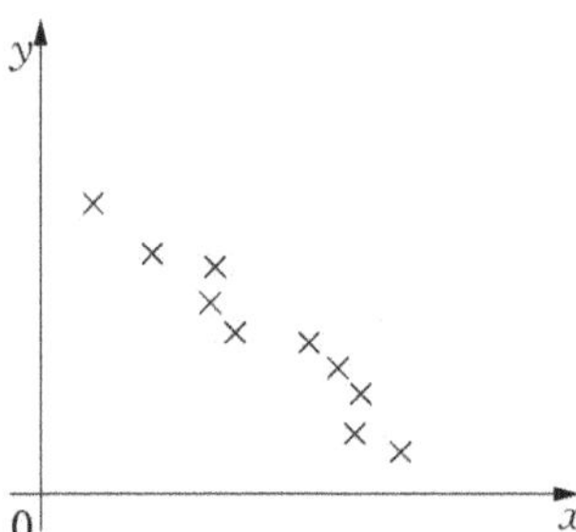

C. Questions that require solutions

8 A study was carried out to investigate the relationship between a group of students' IQs and their performance in a science aptitude test. The total score for the test is 100 marks and the range of IQ values is 80 to 140. The results show that the science aptitude test scores, y, and the IQ values, x, give the line of best fit $y = 2.39 + 0.664x$.

(a) What is the value of the slope of this line of best fit? What does it represent?

(b) What is the value of the y-intercept? Comment on how suitable it is for interpreting someone's science aptitude test score.

(c) Are the test scores and IQ values perfectly positively correlated? Do you think there is a causation between them? Why or why not?

9 The following table shows the marks, x, obtained in a mathematics exam and the marks, y, obtained in a physics exam by a group of eight students.

Mathematics (x)	38	45	52	64	75	78	80	82
Physics (y)	43	60	43	58	78	85	82	72

(a) Using a scale of 2 cm to represent 10 marks on each axis, draw a labelled scatter graph for this set of data.

(b) Find the mean mathematics mark and the mean physics mark.

(c) Plot the mean point on your scatter diagram.

(d) Draw an estimated line of best fit on your scatter diagram and find the equation of the line.

(e) Denny scored 48 marks in the maths exam. He did not sit the physics exam. Estimate Denny's score in the physics exam.

(f) Alice scored 95 marks in the maths exam, but she did not sit the physics exam. State whether it is suitable to use Alice's maths score and the best fit line to estimate her mark in the physics exam. Give a reason for your answer.

Unit test 8

A. Multiple choice questions

1 Look at the table below. To obtain a stratified sample of 100 students, the number of students selected who study maths should be (   ).

Subject	Maths	Physics	Chemistry	English	Total
Number of students	196	165	276	329	866

A. 5 B. 11 C. 19 D. 32

2 Suki claims that using a certain brand of product indicates a higher income. To test her claim, she interviews 50 customers at a shopping centre near her home. The 50 customers she selects form a (   ).

A. convenience sample
B. stratified sample
C. simple random sample
D. cluster sample

3 The time series graph below shows the number of phones sold in a shop in the first six months of a year. The mean number of phones sold is (   ).

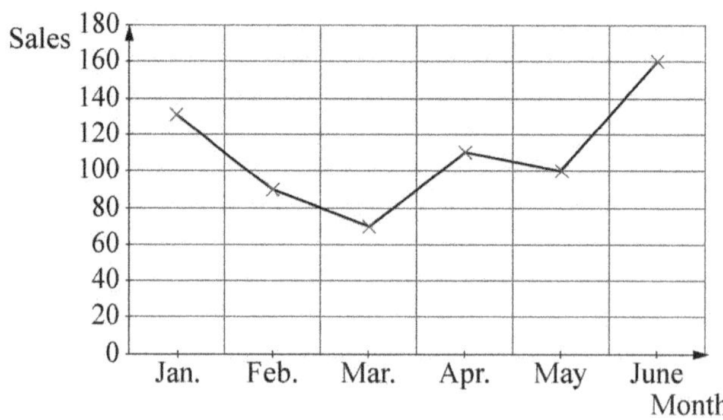

A. 130 B. 110 C. 90 D. 100

4 The measure of the central tendency of a set of data that is most unaffected by extreme values is the (   ).

A. mode B. range C. mean D. median

5 A constant value is added to every data value in a data set. Which of the following statements is correct? (　　)

A. The interquartile range increases by the same constant.

B. The interquartile range increases.

C. The interquartile range is multiplied by the constant.

D. The interquartile range remains the same.

6 Which of the following statements is correct? (　　)

A. A perfectly positive correlation indicates a causation.

B. A perfectly negative correlation indicates a causation.

C. A causation indicates a perfectly positive correlation.

D. A perfectly positive or negative correlation does not necessarily mean a causation.

B. Fill in the blanks

7 Three commonly used measures of central tendency of a set of data are the ________, the ________ and the ________.

8 The following frequency distribution shows the lowest temperatures recorded in 20 cities in a year. The values are given to the nearest degree.

Class (℃)	**Frequency**
1–6	3
7–12	5
13–18	8
19–24	4

To construct a histogram to represent the data, the first bar will begin with the number ________ and end with the number ________. The class width of the second class is ________.

C. Questions that require solutions

9 The maths test results of all Year 11 students in a school are shown in the grouped frequency table below.

Marks	Frequency
41–50	5
51–60	8
61–70	16
71–80	21
81–90	18
91–100	13

(a) How many students are there in Year 11 in the school?

(b) Find the mean test score (to the nearest 1 mark).

(c) Construct a histogram for the data.

(d) Find the class interval that contains the median.

10 A research team plans to conduct a survey of families living in a region to find out how much they spent on Christmas last year. The table below shows the number of the families in each town in the region.

Town	A	B	C	D	E	Total
Number of families	99	139	208	289	325	1060

The team wants to select a sample of these families that best represents the population of the region.

(a) What is the name of the sampling method the team will use? State one advantage of using this method.

(b) The team plans to survey 200 families in total. How many families should the team select from each town?

11 The following frequency table shows the speeds (in km/h) of 105 cars monitored on a motorway.

Speed (km/h)	**Frequency**
$80 \leqslant s < 90$	13
$90 \leqslant s < 95$	8
$95 \leqslant s < 110$	36
$110 \leqslant s < 120$	18
$120 \leqslant s < 140$	30

(a) Find the class interval that contains the median.

(b) Draw a histogram for the information in the table.

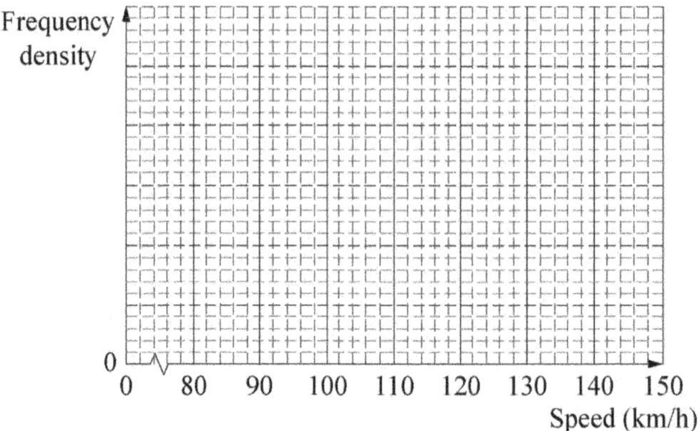

(c) Estimate the number of cars whose speeds are between 90 km/h and 100 km/h.

12 A fitness coach claims that the more time a person spends in a gym, the lower their percentage body mass index. The scatter graph below shows the number of hours each week 10 people spend in a gym and their body mass index.

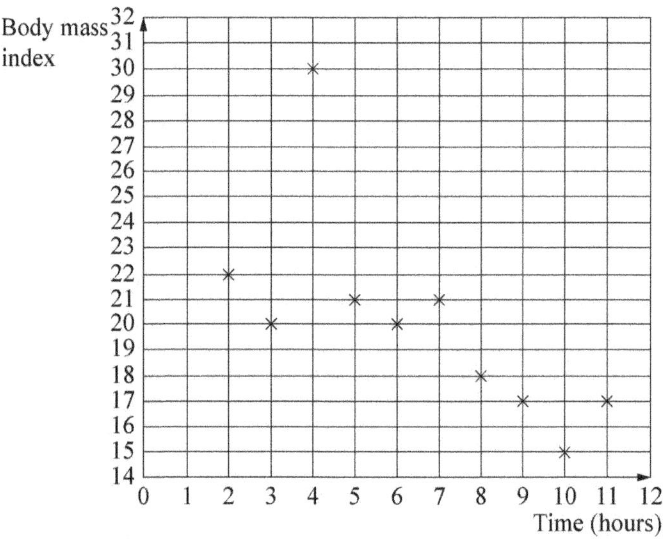

(a) One of the points is an outlier. Write down the coordinates of this point.

(b) Ignoring the outlier, draw a line of best fit on the graph.

(c) Describe the correlation for the data, excluding the outlier.

(d) If a person always spends 6 hours per week in a gym, use the best fit line to predict their body mass index. Comment on the accuracy of your prediction.

End of year test

A. Multiple choice questions (22%)

1 How many of the real numbers -8, 3.1459265, 0, $\frac{\pi}{2}$, $3\sqrt{3}$, $\frac{2}{11}$, $\sqrt{9}$ is/are irrational? ( )

A. one B. two C. three D. four

2 The point $(3, 2)$ is on the graph of the function $y = \frac{k}{x}$. Which of the following statements is incorrect? ( )

A. $k = 6$

B. The graph of the function is in the first and third quadrants.

C. y increases when x increases.

D. When $x < 0$, y decreases as x increases.

3 Given that the quadratic equation $(k - 1)x^2 + 4x - 1 = 0$ has real roots, then the value of k is ( ).

A. $k > -3$ B. $k \geqslant -3$

C. $k > -3$ and $k \neq 1$ D. $k \geqslant -3$ and $k \neq 1$

4 A plan and elevation of a 3D shape is shown in the diagram. This 3D shape could be ( ).

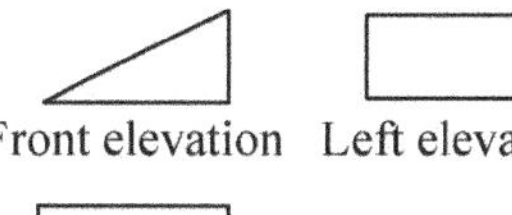

Front elevation Left elevation

Plan

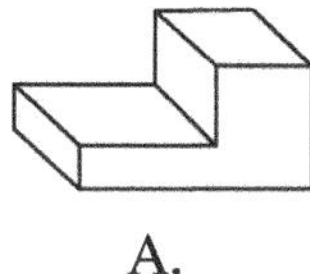

A.

B.

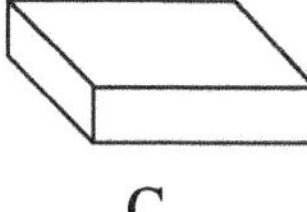

C.

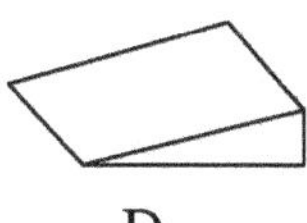

D.

5 The graph of the quadratic function $y = (x + 1)^2 - 2$ is translated with vector $\begin{pmatrix} 2 \\ -3 \end{pmatrix}$. The quadratic function obtained is (   ).

A. $y = (x - 1)^2 - 5$ B. $y = (x - 1)^2 + 1$

C. $y = (x + 3)^2 + 1$ D. $y = (x + 3)^2 - 5$

6 As shown in the diagram, D, E, and F are the midpoints of the three sides of $\triangle ABC$. How many triangles, similar to $\triangle ABC$, are there in the diagram? (   )

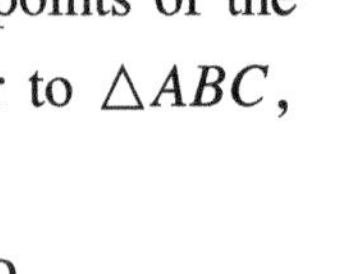

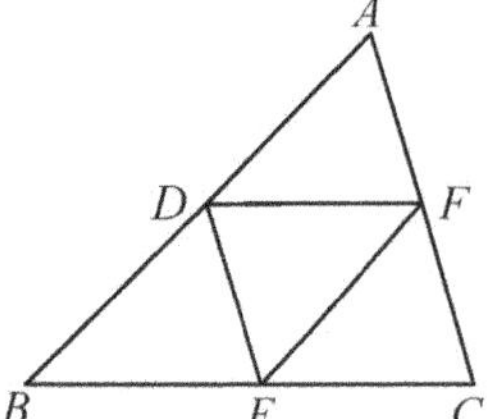

A. one B. two

C. three D. four

7 The ratio of the volume of a cylinder to the volume of a cone is 2 : 3. The radii of the base circles of the cone and the cylinder are the same. The ratio of the height of the cylinder to the height of the cone is (   ).

A. 2 : 3 B. 4 : 5 C. 2 : 1 D. 2 : 9

8 There are 2 English books, 3 maths books and 4 Chinese books on a shelf. A book is selected from the shelf at random. The probability of it being a maths book is (   ).

A. $\frac{2}{9}$ B. $\frac{1}{3}$ C. $\frac{4}{9}$ D. $\frac{5}{9}$

9 Given that in a geometric sequence $\{a_n\}$, if $a_1 a_9 = 9$, then $a_4 a_6$ is (   ).

A. 3 B. ±3 C. 9 D. ±9

10 The centre of a circle is on the y-axis and the radius is 5 units. The equation of the circle passing through the point (−5, 8) is (   ).

A. $x^2 + (y - 8)^2 = 25$ B. $x^2 + (y + 8)^2 = 25$

C. $(x + 5)^2 + (y - 8)^2 = 25$ D. $(x - 8)^2 + y^2 = 2$

11 The equation of a circle is $x^2 + y^2 = 100$. The equation of the tangent to the circle at point (8, 6) is (   ).

A. $4x + 3y = 50$ B. $4x - 3y = 50$ C. $3x + 4y = 48$ D. $3x - 4y = 0$

B. Fill in the blanks (26%)

12 The line $y = 3x - 5$ is reflected in the y-axis and then translated with vector $\begin{pmatrix} 0 \\ -2 \end{pmatrix}$.

The equation of the resulting line is ________.

13 If there are two intersections between the line $y = kx - 6$ and the parabola $y = x^2 + 2x + 3$, then the set of values that k can take is ________.

14 In a right-angled $\triangle ABC$, $\angle C = 90°$ and BD is the angle bisector of $\angle B$. $\triangle BCD$ is folded along line BD so that point C becomes C'. If $AB = 5$ cm and $AC = 4$ cm, then the value of $\sin ADC'$ is ________.

15 The diagram shows the route of a cable car at a local beauty spot. It travels from A to B to C. Using the information given in the diagram, the total length of the driving route is ________.

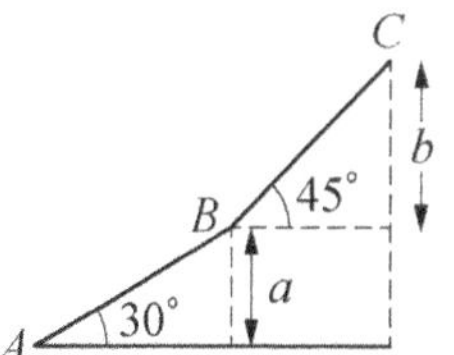

16 If the ratio of the perimeters of two similar triangles is 3 : 4, then the ratio of the areas of the two triangles is ________.

17 The equation of a linear function is $y = (2m - 3)x + 2$. Given that y decreases when x increases, then the set of values that m can take is ________.

18 In $\triangle ABC$, $\angle C = 90°$, $AC = 3$ cm and $BC = 4$ cm. $\triangle ABC$ is rotated around point B so that point A becomes A' and point C becomes C'. Points C', A and B are on the same straight line. Then $\tan AA'C'$ is equal to ________.

19 Given that $f(x) = \sqrt{x}$ and $g(x) = 1 - \cos x$, then the composite function $fg(x) =$ ________.

20 In the right-angled triangle ABC shown in the diagram, $\angle C = 90°$ and ED is perpendicular to BC. $BD = 3$ cm, $DC = 2$ cm and $AB = 6$ cm. Then $BE =$ ________ and $EA =$ ________.

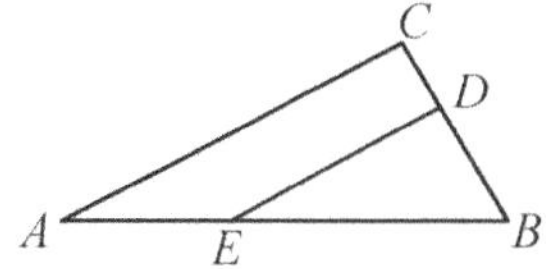

21 If the volume of a sphere is $\frac{9\pi}{2}$ m^3, then the surface area of the sphere is ________ m^2.

22 If the ratio of the areas of two similar triangles is 1 : 2, then the ratio of their corresponding sides is ________.

23 $\{a_n\}$ is an arithmetic progression. Given that $a_1 - a_9 + a_{17} = 7$, then $a_3 + a_{15} =$ ________.

24 The diagram shows a square tablecloth. If a small ball is thrown at random on the tablecloth, the probability of it falling on a shaded part is ________.

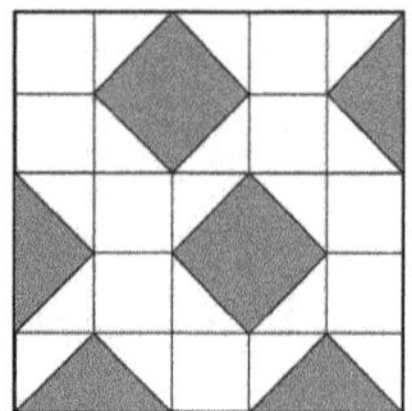

C. Short answer questions (20%)

25 Calculate: $(\pi - 2022)^0 - 1 + \sqrt{2} + \left(\frac{1}{2}\right)^{-2} - 2\sin 45°$

26 Show that the equation $x^3 - x + 1 = 2$ has a solution between $x = 1$ and $x = 2$.

27 $\{a_n\}$ and $\{b_n\}$ are both arithmetic sequences.

(a) For $\{a_n\}$, if the 5th term is 19 and the 8th term is 10, find the 1st term and the common difference.

(b) For $\{b_n\}$, if the 4th term is 10 and the 10th term is 4, find the value of the 14th term.

28 In order to improve the shopping experience of customers in terms of safety and convenience, a shopping centre plans to reduce the slope of an existing escalator. The height AD between two floors is 6 m, the slope angle of the existing escalator (angle ABD) is 30° and the slope angle of the new escalator (angle ACB) is to be 16°, as shown in the diagram. Find how much longer the new escalator needs to be than the existing escalator.

29 The following frequency table shows the masses (in kg) of 50 students in a class.

Mass (kg)	Frequency
$40 \leqslant w < 45$	12
$45 \leqslant w < 50$	8
$50 \leqslant w < 55$	10
$55 \leqslant w < 60$	13
$60 \leqslant w < 65$	7

(a) Find the class interval that contains the median.

(b) Draw a histogram using the information in the table.

(c) Estimate the number of students whose masses are between 48 kg and 53 kg.

D. Questions that require solutions (32%)

30 The graph of the quadratic function $y = ax^2 + bx + c$ intersects the x-axis at points $A(-1, 0)$ and $B(2, 0)$ and intersects the y-axis at point $C(0, 2)$.

(a) Find the equation of the quadratic function.

(b) Find the equation of the line of symmetry and the coordinates of the vertex of the parabola.

31 As shown in the diagram, the graph of the linear function $y = kx + 4$ intersects the x-axis and the y-axis at points A and B, respectively, and $\angle BAO = 45°$.

(a) Find the value of k.

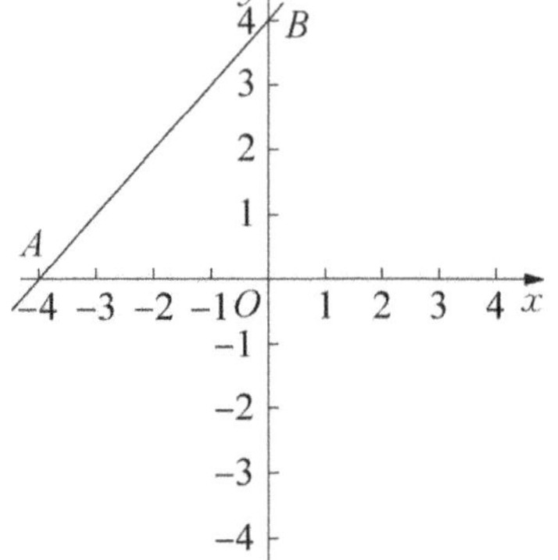

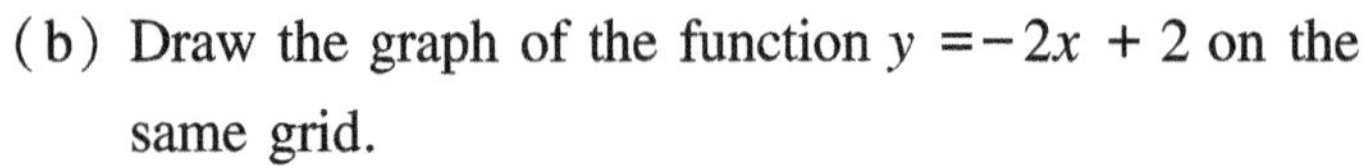

(b) Draw the graph of the function $y = -2x + 2$ on the same grid.

(c) Find the area of the triangle enclosed by the graphs of these two functions and the x-axis.

32 The diagram shows a circle with centre A. AB is the radius of the circle. Point C is on the extension of line AB such that $BC = AB$, and point D is a general point on the circle.

(a) Find the value of $\sin C$ when the area of $\triangle ACD$ is its largest possible value.

(b) What is the value of $\angle A$ when CD is a tangent to the circle? Show your working.

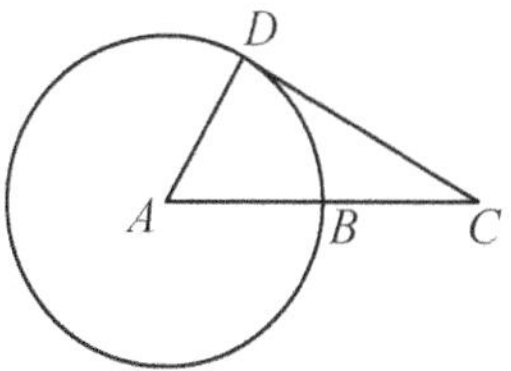

33 A motor insurance company uses a simple random sampling method to select a sample of the insured vehicles that have made a claim on their policies. The amount of compensation that each vehicle in the sample received is shown in the table.

Amount of compensation (£)	0	1000	2000	3000	4000
Number of cars	500	130	100	150	120

(a) Estimate the probability that the amount of compensation is greater than £2000.

(b) Out of the vehicles in the sample, 10% of the insured owners are new drivers. In the sample, 20% of those who received £4000 compensation are new drivers. Considering all the insured vehicles, estimate the probability that a new driver received £4000 compensation.

Answers

Chapter 1 Real numbers

1.1 Calculation with real numbers: Revision

1 A 2 B 3 C 4 B 5 D

6 100 7 $\frac{1}{1006}$ 8 −2 9 $2\sqrt{3}-1$

10 $2\sqrt{2}-4\sqrt{3}$ 11 $11\frac{24}{625}$ 12 (a) $\frac{1}{n}-\frac{1}{n+1}$; (b) $\frac{1}{n(n+1)}=\frac{n+1-n}{n(n+1)}=\frac{n+1}{n(n+1)}-\frac{n}{n(n+1)}=\frac{1}{n}-\frac{1}{n+1}$; (c) $\frac{2009}{2010}$

1.2 Converting between recurring decimals and fractions

1 B 2 D 3 (a) 0.4; (b) $0.\dot{7}1428\dot{5}$; (c) $1.\dot{3}$; (d) $3.\dot{3}$ 4 (a) $\frac{1}{9}$; (b) $\frac{23}{99}$; (c) $\frac{89}{99}$; (d) $\frac{7}{3}$ 5 $\frac{166}{99}$ 6 $0.000\dot{7}8\dot{6}$

7 37 8 (a) $0.\dot{1}+0.\dot{8}$; (b) $0.\dot{7}\times 1.\dot{2}8571\dot{4}$; (c) $0.\dot{8}\div 0.\dot{1}$ (Answers may vary for all of Q8.)

1.3 Rounding and truncating

1 D 2 C 3 A 4 B 5 D

6 (a) 0.00673; (b) 12.58; (c) 5400000; (d) 92300 7 20; 19; 20.0; 19.9 8 2.34; 2 9 3.13; 3 10 8.45; 8.55 11 46 12 22 13 14 14 The smallest possible difference in their masses is 16.5 kg and the largest possible difference is 27.5 kg.

1.4 Solving ratio problems with fractions

1 B 2 A 3 5 : 4 4 5600 5 14 6 28 7 45 8 188 workers are in the first workshop, 150 workers are in the second workshop and 200 workers are in the third workshop.

9 £56, £112, £84, £140 10 160 kilometres 11 40 km/h, 60 km/h

1.5 Solving rate problems with compound units

1 D 2 C 3 10.8, 20 4 1.5, 1000 5 34 6 9 : 8 7 1.67 8 (a) 40 km; (b) 80 km/h 9 0.05 tonnes 10 24 g

1.6 Solving growth and decay problems

1 D 2 A 3 20% 4 25

5 (a) $800000\times(1+3\%)^8=$ £1013416.07; (b) $800000\times(1+3\%)^8-800000=$ £213416.07 6 $200000\times[(1+6\%)+(1+6\%)^2+(1+6\%)^3+(1+6\%)^4+(1+6\%)^5]=$ £1195063.71 7 The deposit: $120\,000\times[(1+5\%)+(1+5\%)^2+(1+5\%)^3+(1+5\%)^4+(1+5\%)^5]=$ £696229.54, the interest: $696229.54-120000(5)=$ £96229.54

8 2130000 9 (a) $w=500(1-10\%)^t=500\times 0.9^t$; (b)

t	1	2	3	4	5	6	7	8	9	10
w	450	405	364.5	328.05	295.245	265.7205	239.14845	215.233605	193.7102445	174.33922005

(c) 6–7 years

1.7 Solving equations numerically using iteration

1 B 2 C 3 B 4 5, 7, 11, 19 (Hint: answer may vary.) 5 88000, 112312

6 When $x=1$, $x^5-4x^2-1=-4<0$; when $x=2$, $x^5-4x^2-1=15>0$, so the equation $x^5-4x^2-1=0$ has a solution between $x=1$ and $x=2$.

7 When $x=-1.9$, $x^3+2x-(-10)=-0.659<0$; when $x=-1.8$, $x^3+2x-(-10)=0.568>0$, so the equation $x^3+2x=10$ has a solution between $x=-1.9$ and $x=-1.8$. 8 Since the equation $3x^3+2x-1=0$ can be rearranged to give $3x^3=1-$

$2x$, then $x = \sqrt[3]{\frac{1-2x}{3}}$ **9** 1.369 **10** 0.74

11 1.4 **12** (a) When $x = 2$, $x^3 - 5x + 1 = -1 < 0$, when $x = 2.2$, $x^3 - 5x + 1 = 0.648 > 0$, so the equation has a root between $x = 2$ and $x = 2.2$. (b) Since the equation $x^3 - 5x + 1 = 0$ can be rearranged to give $x^3 = 5x - 1$, then $x = \sqrt[3]{5x-1}$. (c) 2.1

Unit test 1

1 B **2** C **3** B **4** C **5** A **6** B

7 C **8** A **9** 8.85×10^3 **10** 4, 5

11 $\frac{\sqrt{3}}{2} + \frac{1}{2}$ **12** $\sqrt{2}$, $\sqrt{3}$ (Answers may vary.)

13 20, 22 **14** 1.52 **15** 17 **16** 3 : 4

17 1.3 **18** (a) £60; (b) $\frac{16}{3}$ km/h; (c) $\frac{32}{3}$ km **19** (a) 1.5 hours; (b) 4:30 p.m.

20 (a) 90; (b) 62.5; (c) 240

21 (a) $\sqrt{n+1} - \sqrt{n}$, $\sqrt{n+1} + \sqrt{n}$

(b) $\frac{1}{\sqrt{n+1}+\sqrt{n}} = \frac{1 \times (\sqrt{n+1}-\sqrt{n})}{(\sqrt{n+1}+\sqrt{n})(\sqrt{n+1}-\sqrt{n})} = \sqrt{n+1} - \sqrt{n}$, $\frac{1}{\sqrt{n+1}-\sqrt{n}} = \frac{1 \times (\sqrt{n+1}+\sqrt{n})}{(\sqrt{n+1}-\sqrt{n})(\sqrt{n+1}+\sqrt{n})} = \sqrt{n+1} + \sqrt{n}$; (c) $\sqrt{2022} - 1$ **22** (a) When $x = 1$, $x^3 - 5x + 3 = -1 < 0$, when $x = 2$, $x^3 - 5x + 3 = 1 > 0$, so the equation has a root between $x = 1$ and $x = 2$. (b) Since the equation $x^3 - 5x + 3 = 0$ can be rearranged to give $x^3 = 5x - 3$, then $x = \sqrt[3]{5x-3}$. (c) 1.83

Chapter 2 Function and graphs (I)

2.1 Review of functions

1 A **2** D **3** A **4** $m < 3$ **5** $a = -3$

6 $0 \leqslant m \leqslant 1$ **7** $\left(\frac{1}{3}, 0\right)$ **8** $y = x - 8$

9 (a) $m = 3$, $y = \frac{3}{4}x$ (b) (10, 0) or (−10, 0) **10** (a) $y = 1.8x + 32$ (b) 30℃

2.2 Inputs and outputs, inverse functions and composite functions

1 D **2** C **3** D **4** C **5** $x = \frac{1}{2}y + \frac{3}{2}$ **6** $y = -\frac{6}{x}$ **7** (1, 0) **8** ±2

9 $y = -(x+2)^2 + 3$; $x_1 + x_2 = -4$ **10** 10, 16, 10, 16 **11** $y = -3x - 3$; No **12** (a) $y = -\frac{8}{x}$

(b) (i) $y = -\frac{8}{x} + 1$, (ii) $y = -\frac{8}{x-1}$.

13 (a) (i) $a = \frac{1}{4}$, $b = \frac{1}{2}$ (ii) $m = \frac{2}{3}$, $n = 3$ (iii) $p = -\frac{2}{11}$, $q = -\frac{11}{2}$ (b) $y = 2x$

(c) $y = \frac{2}{x}$

2.3 Further work about linear functions (1)

1 B **2** B **3** D **4** A **5** $\frac{2}{3}$

6 $y + 1 = 3(x - 2)$, i.e. $y = 3x - 7$

7 (−2, 4) **8** $y = 3x - 2$ (Answers may vary.) **9** (3, −2) **10** $y = -4x$ or $y = -x + 3$

11 (−1, −1) **12** Diagram omitted. The solution for the system of equations is $\begin{cases} x = -1 \\ y = 2 \end{cases}$.

13 Gradient of AB = gradient of $AC = 2$; a shared point and the same gradient mean all 3 points are on a straight line.

2.4 Further work about linear functions (2)

1 A **2** C **3** D **4** D **5** $y = -\frac{1}{2}x + 2$ **6** 3 **7** $3x + 4y - 14 = 0$ **8** $y = \frac{3}{2}x + 3$ **9** (1, −3) **10** $0 < k < 2$ **11** $y = -x + 5$ **12** The solution to the system is $\begin{cases} x = 5 \\ y = -\frac{3}{2} \end{cases}$.

13 (a) $y = -x - 1$ (b) When the point of intersection is in the first quadrant, $k > 1$. When the point of intersection is in the second quadrant, $k < -1$; When the point of intersection is in the fourth quadrant, $-1 < k < 1$.

2.5 Graphs of functions (1)

1 C 2 D 3 C 4 C 5 2 6 $x + 2y - 4 = 0$ 7 1 8 $\frac{3}{2}$ 9 $x < -\frac{1}{2}$ or $x > 2$ 10 $x < 1$ 11 Diagram omitted. 12 $x_1 = 1$, $x_2 = x_3 = -2$ 13 3

2.6 Graphs of functions (2)

1 D 2 C 3 B 4 D 5 (−2, 4) 6 $y = -\frac{6}{x}$ 7 3 8 $1 < x < 2$ or $x < 0$ 9 2 10 $c < b < a$ 11 (a) $y_1 = -x + 6$, $y_2 = \frac{5}{x}$ (b) $0 < x < 1$ or $x > 5$ 12 5 13 3

Unit test 2

1 D 2 D 3 B 4 C 5 B 6 A 7 $y = -x$ 8 $\pm\sqrt{3}$ 9 $y = -1$ 10 $3x + 2y - 3 = 0$ 11 $y = -\frac{4}{3}x + 2$ 12 16, 9, 7, 25 13 $3x + y - 13 = 0$ 14 $\left(\frac{4}{7}, -\frac{1}{7}\right)$ 15 $m > -1$ 16 1 17 (a) $y = -\frac{1}{2}x + \frac{1}{2}$; (b) (−3, 2); (c) $-\frac{1}{2}$; (d) 2 18 (a) $\sqrt{2}$, 5; (b) $2\sqrt{2} + 3$; (c) $y = \frac{x-3}{2}$; (d) $-\frac{1}{2}$ or $-\frac{5}{2}$ 19 (a) (0, 2); (b) $y = 2x + 2$

Chapter 3 Function and graphs (II)

3.1 Translations and reflections of the graph of a given function

1 D 2 B 3 D 4 D 5 $y = \frac{1}{2}x - 3$ 6 2, left, $y = 2x$ 7 $y = \frac{1}{2}x^2 - x + 3$ 8 $-\frac{2}{3}$ 9 $y = -3x - 2$ 10 (a) the graph is symmetrical about the y-axis; (b) $x \neq 0$. (Answers may vary: e.g. $y > 0$.) 11 (a) $y = -\frac{3}{2}x + 6$ (b) $k \neq \pm\frac{3}{2}$ 12 $y = 3x + 9$ 13 Graph is omitted. When $k < -4$, there are no solutions; when $k > -4$, there are two solutions; when $k = 0$, there is one solution.

3.2 Graphs of quadratic functions $y = a(x + m)^2 + k$ (1)

1 D 2 A 3 D 4 B 5 $x = -1$ 6 (6, −3) 7 6, minimum, −3 8 (1, 2) 9 $y = (x - 1)^2 - 1$ 10 (−4, −4) 11 The parabola is an inverted U-shape, $x = 1$, (1, 8); graph omitted. 12 (a) $y = -2x^2 - 4x + 4$ (b) $y = -2(x + 1)^2 + 6$, (−1, 6) 13 $y = 2x^2 - 8x + 6$, (2, −2)

3.3 Graphs of quadratic functions $y = a(x + m)^2 + k$ (2)

1 D 2 B 3 D 4 B 5 (0, 3) 6 $m > \frac{1}{4}$ 7 decreases, increases 8 $x = 2$ 9 third 10 0 11 $y = -\frac{1}{2}(x + 3)^2 + 2$; (a) (−3, 2), $x = -3$, graph omitted; (b) $-5 < x < -1$ 12 $y = -x^2 + 4x - 2$ 13 (a) $\Delta = m^2 - 4(m - 5) = (m - 2)^2 + 16 > 0$, so the parabola always has two points of intersection with the x-axis for whatever value that m takes. (b) $m = 2$

3.4 Graphs of quadratic functions $y = a(x + m)^2 + k$ (3)

1 A 2 C 3 B 4 C 5 −1 6 (1, 2) 7 $y = -4(x - 2)^2 + 3$ 8 ±4 9 > 10 $y = -x + 2$ 11 (a) $b = -7$, $c = 12$ (b) 4 12 (a) $y = x^2 + 4x + 3$ (b) $y = x^2 - 2x - 1$ 13 (a) There are two intersection points of the parabola and the line whatever values m takes. (b) $\frac{3\sqrt{5}}{5}$

3.5 Graphs of quadratic functions $y = a(x+m)^2 + k$ (4)

1 B 2 A 3 B 4 D 5 $a < 2$ 6 third 7 2 8 $y = (x+1)^2 - 4$ (Answers may vary.) 9 $y = -\sqrt{3}(x^2 - 4x + 3)$ 10 $\frac{1 \pm \sqrt{5}}{2}$ or $m = 0$ 11 (a) $y = x^2 - 5x + 7$, $\left(\frac{5}{2}, \frac{3}{4}\right)$, $(0, 7)$; (b) $2 < x < 3$ 12 (a) $y = -x^2 + 4x + 5$; (b) yes, $t = -4$ 13 (a) $y = x^2 + 4x + 4$ (b) $\left(-\frac{9}{2}, \frac{25}{4}\right)$

3.6 Graphs of other functions

1 C 2 C  3 A 4 C 5 20 6 0.8 7 1 8 37 800 9 down, $t = 0$, 12 10 4 11 (a) 0.5 hours; (b) 20 km/h, 60 km/h; (c) 20 km 12 (a) 1 hour, 200 mg/100 ml. (b) 225 (c) Above. When $x = 2.5$, $y = \frac{225}{2.5} = 90 > 80$. 13 6. (b) The distance moved by the particle is equal to the area of the trapezium $OABH$.

3.7 Gradient at a point on a curve

1 C 2 A  3 C 4 B 5 2 6 $y = -3x$ 7 $k > 0$ 8 decreases 9 $\bar{v}_1 = 41$ m/s, $\bar{v}_2 = 59$ m/s 10 $\frac{f(x_0 + m) - f(x_0)}{(x_0 + m) - x_0} = -\frac{2}{x_0(x_0 + m)}$ 11 (a) $-12 - 6m - m^2$ (b) The slope of the tangent line l is $k = -12$; the equation that defines it is $y + 8 = -12(x - 2)$, i.e. $y = -12x + 16$.

Unit test 3

1 B 2 B 3 C 4 B 5 B 6 A 7 B 8 $y = -3x - 7$ 9 $y = x^2 + 2x + 3$ 10 (a) The graph is symmetrical about the line $x = 1$; (b) When $x < 1$, the value of y decreases as the value of x increases. (Answers may vary.) 11 $3x + 2y - 3 = 0$ 12 k is any real number 13 $x = \frac{3}{8}$ 14 $y = -(x-1)^2 + 4$ (Answers may vary.) 15 $x > -\frac{3}{4}$, $x < -\frac{3}{4}$ 16 1.6 17 (a) When $a > -4$, there are 2 solutions; when $a = -4$, there is one solution; when $a < -4$, there is no solution. 18 (a) Graph is symmetrical about $x = -1$ and crosses x-axis at -3. (b) $y = -x^2 - 2x + 3$ 19 (a) $x^2 - x + 3 = 3x - a \rightarrow x^2 - 4x + (3 + a) = 0$ (b) $x = 5$ (c) $a = 1$ 20 (a) $\frac{\sqrt{2}}{2}$; (b) $\frac{3}{\sqrt{2}}$ 21 (a) $m = 2$, $m = -3$ (b) $m = -3$, $(4, 15)$; (c) when $m = -3$, $x > 4$; when $m = 2$, $x < -1$ 22 (a) 8.2; (b) 2.24; (c) Answer omitted.

Chapter 4 Trigonometric ratios and trigonometric functions

4.1 Evaluating trigonometric ratios of acute angles (1)

1 B 2 C 3 D 4 D 5 $\sqrt{3} + 1$ 6 $\frac{\sqrt{3}}{2}$ 7 $\frac{5}{12}$ 8 $2\sqrt{6}$ 9 $\frac{3}{5}$ 10 (a) 1 (b) $\frac{5}{6}$ 11 $\frac{\sqrt{2}}{2}$ 12 $8\sqrt{3}$

4.2 Evaluating trigonometric ratios of acute angles (2)

1 B 2 C 3 B 4 B 5 75° 6 15° 7 30° 8 $\frac{\sqrt{2}}{4}$ 9 6 10 6 11 $\angle D = 15°$, $\sin 15° = \frac{\sqrt{6} - \sqrt{2}}{4}$, $\cos 15° = \frac{\sqrt{6} + \sqrt{2}}{4}$, $\tan 15° = 2 - \sqrt{3}$ 12 (a) $\sqrt{5}$ (b) $\frac{4}{5}$

4.3 Angles of any size and their measures (1)

1 D 2 D 3 D 4 C 5 $k \times 360° + 120° (k \in \mathbf{Z})$ 6 third 7 fourth 8 315° 9 (a) $k \times 360° (k \in \mathbf{Z})$ (b) $k \times 360° + 180° (k \in \mathbf{Z})$ (c) $k \times 360° + 90° (k \in \mathbf{Z})$ (d) $k \times 360° + 270° (k \in \mathbf{Z})$ 10 (a) 80°, first quadrant (b) 320°, fourth quadrant

11 (a) $-345°$ (b) $600°$ (c) $-540°$ (d) $40°$ 12 $45° + k \times 180° (k \in \mathbf{Z})$

4.4 Angles of any size and their measures (2)

1 A 2 D 3 $-90°$ 4 $-360° + 252°$, $-360° + 135°$ (Answers may vary.) 5 2
6 third, $230°$, $-130°$ 7 second or fourth
8 5 9 (a) $k \times 180° (k \in \mathbf{Z})$ (b) $k \times 180° + 90° (k \in \mathbf{Z})$ (c) $k \times 90° (k \in \mathbf{Z})$
10 (a) $15°$, $180°$, $270°$, $120°$, $1200°$ and $460°$ (b) $-100°$, $-360°$, $-200°$ and $-180°$ (c) $180°$, $-360°$, $0°$ and $-180°$ (d) $120°$, $-200°$, $1200°$ and $460°$ 11 (a) in the second or fourth quadrant (b) $k \times 120° + 90° < \frac{\alpha}{3} < k \times 120° + 120° (k \in \mathbf{Z})$ (c) $k \times 720° + 540° < 2\alpha < k \times 720° + 720° (k \in \mathbf{Z})$ 12 $\alpha = 65°$, $\beta = 15°$

4.5 Sine rule, cosine rule and solving scalene triangles (1)

1 B 2 C 3 C 4 $\frac{\sqrt{3}}{2}$ 5 13 cm
6 $5\cos\alpha$ 7 B 8 D 9 $\frac{4}{5}$, $\frac{3}{5}$, $\frac{4}{3}$
10 $\frac{3\sqrt{3}}{2}$ 11 $\angle A = 30°$, $\angle B = 60°$, $a = 5$, $b = 5\sqrt{3}$

4.6 Sine rule, cosine rule and solving scalene triangles (2)

1 C 2 B 3 C 4 C 5 $\frac{3}{4}$
6 $\sqrt{2}$ 7 $\sqrt{17}$ 8 $\sqrt{3} - 1$ 9 $CD = 3$ cm 10 $12\sqrt{3}$, 24 11 (a) 27 (b) $\frac{\sqrt{5}}{5}$

4.7 Sine rule, cosine rule and solving scalene triangles (3)

1 A 2 D 3 B 4 D 5 $\frac{2\sqrt{2}}{3}$
6 $2\sqrt{7}$ 7 $\frac{1}{7}$ 8 30 9 $4\sqrt{5}$ 10 $\frac{3}{4}$
11 (a) 5 (b) $\frac{\sqrt{2}}{10}$

4.8 Sine rule, cosine rule and solving scalene triangles (4)

1 D 2 D 3 $1 : \sqrt{3} : 2$ 4 $30°$ 5 $\frac{2\sqrt{2}}{3}$
6 $2\sqrt{37}$ 7 $90°$ 8 8 9 $\frac{3\sqrt{3}}{2}$ 10 8
11 An isosceles triangle or a right-angled triangle
12 $\sqrt{21}$ or $\sqrt{61}$ 13 $c = \sqrt{2} + \sqrt{6}$, $\angle B = 60°$, $\angle C = 75°$

4.9 Trigonometric functions (1)

1 A 2 A 3 C 4 $-1 \leqslant y \leqslant 3$
5 $(k \times 180°, 0) (k \in \mathbf{Z})$ 6 $\pm\frac{1}{2}$, ± 1
7 $-1 \leqslant y \leqslant 1$ 8 Answers omitted.
9 The graphs are omitted. The graph of $y = \cos x (-90° \leqslant x \leqslant 270°)$ can be translated right by $90°$ and up by 1 unit. 10 (a) $-180° \leqslant x < -150°$ or $-30° < x \leqslant 180°$ (b) $-180° \leqslant x \leqslant 30°$ or $150° \leqslant x \leqslant 180°$

4.10 Trigonometric functions (2)

1 D 2 B 3 D 4 1 5 $180° + k \times 360° (k \in \mathbf{Z})$ 6 $270° + k \times 360° (k \in \mathbf{Z})$
7 $a = \frac{1}{2}$, $b = \pm 1$ 8 $\frac{1}{4}$, -2 9 6, $\frac{1}{4}$
10 $a > 0$ or $a < -1$

Unit test 4

1 C 2 D 3 A 4 C 5 D 6 A
7 B 8 D 9 D 10 A 11 3 12 $\frac{7}{5}$
13 second, $120°$, $-240°$ 14 $\frac{3}{5}$ 15 $8\sqrt{5}$
16 $-\frac{\sqrt{2}}{2} \leqslant x \leqslant \frac{\sqrt{3}}{2}$ 17 ± 1 18 $-\frac{1}{2} \leqslant x \leqslant 1$, $\frac{\pi}{3} \leqslant x \leqslant \pi$ 19 2.4 20 $1.6 + \frac{13\tan 35°}{1 - \tan 35°}$
21 (a) $\sqrt{5}$ (b) $\frac{5\sqrt{3}}{6}$ 22 (a) Let $\angle BCD = x = \angle BDC$; $\angle DBC = 180 - 2x$; $\angle ABD = 2x -$

90; $\angle BCE = 2x - 90$; then $BE = BC\sin(2x - 90)$ and $AD = BD\sin(2x - 90)$; but $BD = BC$ (given), so $BE = AD$. (b) $4 + 2\sqrt{3}$ **23** (a) Graph omitted. (b) The graph of $f(x) = \sin x (0° \leqslant x \leqslant 360°)$ can be translated down by 2 units. (c) There are no such values. (d) $\frac{9}{4}$, 2

Chapter 5 Sequences

5.1 Revision of sequences

1 B **2** B **3** C **4** D **5** 2, 4, 6, 8, 10 **6** Diagram omitted, $\frac{n(n+1)}{2}$; diagram omitted, $4n$ **7** –4 **8** $\pm\frac{1}{2}$

9 $3 \times \left(\frac{1}{2}\right)^{n-1}$ **10** fifth **11** 12 **12** (a) $\frac{28}{31}$. (b) No, it isn't. (c) $a_n = \frac{3n-2}{3n+1} = 1 - \frac{3}{3n+1}$, $0 < \frac{3}{3n+1} < 1$, so $0 < a_n < 1$. (d) Yes, there is one term.

5.2 Arithmetic sequences

1 C **2** B **3** B **4** $2n - 12$ **5** –11 **6** 24 **7** –5, –88, 22nd **8** 13 **9** $9n - 8$ **10** 53 **11** $b_{n+1} - b_n = (a_{2n+1} + a_{2n+2}) - (a_{2n-1} + a_{2n}) = (a_{2n+1} - a_{2n-1}) + (a_{2n+2} - a_{2n}) = 4d$, so the sequence $\{b_n\}$ is also an arithmetic sequence and the common difference of $\{b_n\}$ is 4d. **12** 18

13

6	4	2	0	–2
11	9	7	5	
16	14	12		
21	19			
26				

5.3 Geometric sequences

1 C **2** A **3** D **4** $3\left(-\frac{1}{3}\right)^{n-1}$

5 $\left(\frac{1}{7}\right)^{n-7}$ **6** $\frac{1}{2}$ **7** 2, 1 **8** 4

9 56 or –24 **10** $a = 0$ **11** (a) $a_n = a_1q^{n-1}$, then $a_{2k+1} = a_1q^{2k}$, so $\frac{a_{2(k+1)+1}}{a_{2k+1}} = q^2$, so the sequence $\{a_{2k+1}\}$ (k is a natural number) is a geometric sequence. (b) $\sqrt{a_n} = \sqrt{a_1q^{n-1}}$, so $\frac{\sqrt{a_{n+1}}}{\sqrt{a_n}} = \sqrt{q}$, so the sequence $\{\sqrt{a_n}\}$ is a geometric sequence.

12 $\frac{13}{16}$ **13** 9

5.4 Fibonacci sequences

1 A **2** B **3** C **4** B **5** 34 **6** 13 **7** 1, 2, $\frac{3}{2}$, $\frac{5}{3}$, $\frac{8}{5}$ **8** 2020th **9** $a - 1$ **10** $1 - m$

11

$$\begin{aligned} a_{n+1} + a_n &= \frac{\alpha^{n+1} - \beta^{n+1}}{\alpha - \beta} + \frac{\alpha^n - \beta^n}{\alpha - \beta} \\ &= \frac{\alpha^{n+1} + \alpha^n}{\alpha - \beta} - \frac{\beta^{n+1} + \beta^n}{\alpha - \beta} \\ &= \frac{\alpha^n(\alpha + 1)}{\alpha - \beta} - \frac{\beta^n(\beta + 1)}{\alpha - \beta} \\ &= \frac{\alpha^n \times \alpha^2}{\alpha - \beta} - \frac{\beta^n \times \beta^2}{\alpha - \beta} \\ &= \frac{\alpha^{n+2}}{\alpha - \beta} - \frac{\beta^{n+2}}{\alpha - \beta} \\ &= a_{n+2} \end{aligned}$$

5.5 More about sequences

1 D **2** C **3** C **4** D **5** 64, 512 **6** 55 **7** triangular **8** 150 **9** decreases **10** (a) 55 (b) $a_n = \frac{n(n+1)}{2}$ (c) $b_{2k-1} = \frac{(5k-1)5k}{2}$, $b_{2k} = \frac{5k(5k+1)}{2} (k \in \mathbf{Z}, k \geqslant 1)$, $b_{2019} = 12748725$, 5049th **11** $\frac{1}{16}$ **12** From the given, we get $n^2 + n^3 + \frac{n(n+1)}{2} = n(n+1)\left(n + \frac{1}{2}\right)$, which is divisible by $(n+1)$ when n is an even number.

5.6 Calculate the *n*th term of linear and quadratic sequences

(1) C (2) D (3) C (4) C (5) $5n-10$ (6) 4 (7) $6n-3$ (8) 172 (9) n^2-1 (10) $3n-21$ (11) $b_n=2n-1$, $b_n=4n-3$ or $b_n=8n-7$ (12) (a) $7-2\neq 14-7$, therefore it is not part of a linear sequence. (b) Since the expression of the quadratic sequence n^2+2n-1 can represent the first six terms of the sequence. (c) n^2+2n-1

Unit test 5

(1) A (2) C (3) C (4) B (5) A (6) C (7) C (8) C (9) C (10) B (11) 591, 82nd (12) -5 (13) $2^{\frac{n+1}{2}}$ (14) 288th (15) 280 (16) 253, $\frac{n(n+1)}{2}$ (17) 1, 2, $\frac{5}{2}$, $\frac{29}{10}$ (18) 361 (19) 3rd (20) infinitely many (21) 13 (22) (a) Because $-1-(-6)=4-(-1)=9-4=5$, the four numbers can form the first four terms of an arithmetic sequence. (b) $-6+5(n-1)$. (c) 489; (d) 102nd (23) (a) Because $\frac{1}{a_{n+1}+1}-\frac{1}{a_n+1}=\frac{a_n+2}{a_n+1}-\frac{1}{a_n+1}=1$, $\left\{\frac{1}{a_n+1}\right\}$ is an arithmetic sequence. (b) $\frac{1}{a_n+1}=\frac{1}{a_1+1}+1\times(n-1)=\frac{2n-1}{2}$, so $a_n=\frac{3-2n}{2n-1}$ (24) (a) 15, 25 (b) $\frac{(n+1)(n+2)}{2}$, $(n+1)^2$ (c) Let $\frac{(n+1)(n+2)}{2}$ be 36, then $n=7$. Let $(n+1)^2$ be 36, then $n=5$, so 36 is both a triangular number and a square number. (d) 1 and 1225

Chapter 6 Two- and three-dimensional shapes

6.1 Enlargement of fractional and negative scale factors

 B  B (3) D (4) C (5) A (6) 3 (7) 100, $\sqrt{10}$ (8) $\frac{21}{2}$ (9) $\frac{2}{3}$ (10) $10\sqrt{5}-10$ (11) 9 (12) (a) $A'(3,-3)$, $B'(-3,7.5)$, $C'(-6,3)$ (b) $A''(-4,4)$, $B''(4,-10)$, $C''(8,-4)$ (13) 150 cm^2, 12 cm (14) 6 km (15) 10 cm

6.2 Combined transformation of geometric shapes

(1) B (2) D (3) C (4) A (5) B (6) 5:30 (7) 2 cm, point C, CF, $\angle ECF$ (8) 10 (9) 74° (10) $\frac{169}{4}\pi+30$ (11) (a) new position is $A'(0,3)$, $B'(-2,5)$, $C'(-5,2)$ (b) new postion is $A''(0,0)$, $B''(2,-2)$, $C''(-1,-5)$; Size is the same, shape is the same, position and orientation have changed. (12) (a) Point A, 90° (b) $\triangle AEF$ is a right-angled isosceles triangle, Area $=\frac{17}{32}$ (Hint: $\text{area}_{\triangle AEF}=\text{area}_{\text{quadrilateral } AFCE}-\text{area}_{\triangle EFC}=\text{area}_{\text{quadrilateral } ABCD}-\text{area}_{\triangle EFC}$.) (13) (a) 60°, anticlockwise (b) There are three pairs: $\triangle BCD$ and $\triangle ACE$, $\triangle BCG$ and $\triangle ACF$, $\triangle DGC$ and $\triangle EFC$ (c) 80° (14) (a) Diagram omitted. (b) 62° (c) 16 (Hint: $\frac{\text{area}_{\triangle BDE}}{\text{area}_{\triangle DEC}}=\frac{BE}{EC}=\frac{1}{3}$, $\text{area}_{\triangle BDE}=\text{area}_{\triangle DEF}=2$.)

6.3 Equation of a circle

(1) C (2) D (3) B (4) D (5) B (6) $(x-1)^2+(y-2)^2=5$ (7) $(x-3)^2+(y+4)^2=1$ (8) $\frac{1}{3}$ (9) $x^2+(y-3)^2=1$ (10) $(x-3)^2+(y+4)^2=25$ (11) Let the equation of the circle be $(x-a)^2+(y-b)^2=r^2$. The centre of the circle is on $y=0$ and $b=0$. Since the circle passes through the points $A(1,4)$ and $B(3,2)$, we get $(1+a)^2+16=r^2$ and $(3-a)^2+4=r^2$. So the solutions to the system of equations are: $a=-1$ and $r^2=20$. Hence the equation to the circle is $(x+1)^2+y^2=20$. Since the distance from point $P(2,4)$ to the centre

of the circle $C(-1, 0)$ is $d = |PC| = \sqrt{(2+1)^2 + 4^2} = \sqrt{25} > r$, point P is outside the circle. 12 Since point $P(1, -\sqrt{3})$ is on the circle with centre O, the equation of the tangent line PT can be written as $y = k(x-1) - \sqrt{3}$. From $d = r$, and $\frac{-k-\sqrt{3}}{\sqrt{1+k^2}} = 2$, so the solution is $k = \frac{\sqrt{3}}{3}$. Therefore, $y = \frac{\sqrt{3}}{3}(x-1) - \sqrt{3}$. 13 (a) $r^2 = (x-k)^2 + (x-k)^2 = (k-6)^2 + (k-4)^2 = 2k^2 - 20k + 52$ (b) $r^2 = 2k^2 - 20k + 52 = 2(k-5)^2 + 2$, so when $k = 5$, r is minimum. The minimum value is $\sqrt{2}$. In this case, the standard form of the equation of the circle is $(x-5)^2 + (y-5)^2 = 2$. 14 $(x-4)^2 + (y-5)^2 = 16$

6.4 Bearings

1 C 2 C 3 B 4 C 5 $2\sqrt{3}$ 6 60sin 10° 7 035°, 315°, 8 100tan 45° cos 35° + 100sin 35° = 139 m (3 s.f.)

9 (a) 176.31°, (b) $\sqrt{13}$ km 10 (a) 075°, (b) 60cos 15° = 58.0 km 11 (a) Construct BF perpendicular to l intersecting at F. In the right-angled triangle ADE, $\angle DAE = 60°$, $AD = 2$. So $AE = 4$, $DE = 2\sqrt{3}$. Since $AB = 10$, $BE = 6$. In the right-angled triangle BFE, $\angle BEF = 30°$, so $BF = 3$ km (b) $EF = 3\sqrt{3}$, $DF = DE + EF = 5\sqrt{3}$. In the right-angled triangle BFC, $\angle CBF = 76°$. So $CF = BF\tan 76° = 3\tan 76°$, $CD = CF - DF = 3\tan 76° - 5\sqrt{3}$, 5 min $= \frac{1}{12}$ hour. Therefore, the speed of the ship is $12(3\tan 76° - 5\sqrt{3}) = 40.5$ km/h (3 s.f.).

6.5 Introduction to plans and elevations of 3D shapes

1 B 2 C 3 D 4 D 5 ①②③④ 6 $\frac{5}{2}\sqrt{\pi^2 + 4}$ cm 7 $\sqrt{11}$ 8 2 9 $2 + \sqrt{3}$ 10 30π 11 $2\sqrt{5}$ 12 (a) Unfold the lateral surface of the cone on a plane by cutting along SA, as shown in the diagram, then the unfolded shape is a sector, and the length L of the arc AA' is the perimeter of the base of the cone, so $L = 2\pi r = 2\pi$, thus $\angle ASM = \frac{L}{2\pi l} \times 360° = \frac{2\pi}{2\pi \times 4} \times 360° = 90°$. The minimum length of the rope is the length of AM in the unfolded shape. Since $\angle ASM = 90°$, using Pythagoras' theorem, $AM = \sqrt{x^2 + 16}$ $(0 \leqslant x \leqslant 4)$. Therefore, $f(x) = AM^2 = x^2 + 16$ $(0 \leqslant x \leqslant 4)$ (b) When the rope is the shortest, in the unfolded shape, construct SR perpendicular to AM with R being the foot of the perpendicular. Then the length of SR is the shortest distance from S to the rope. In $\triangle SAM$, $\text{area}_{\triangle SAM} = \frac{1}{2}SA \times SM = \frac{1}{2}AM \times SR$. Thus, $SR = \frac{SA \times SM}{AM} = \frac{4x}{\sqrt{x^2+16}}$ $(0 \leqslant x \leqslant 4)$, that is, when the rope is the shortest, the distance from the apex to the rope is $\frac{4x}{\sqrt{x^2+16}}$ $(0 \leqslant x \leqslant 4)$.

6.6 Surface areas of 3D shapes

1 C 2 B 3 C 4 96π 5 180 6 90π 7 8π 8 $\sqrt{S_1} + 2\sqrt{S_2} = 3\sqrt{S_3}$ 9 3 : 16 10 In Rt$\triangle ACB$, $AC = BC = 5$; then $AB = 5\sqrt{2}$. In Rt$\triangle SAB$, $SB = 5\sqrt{5}$, $AB = 5\sqrt{2}$, then $SA = 5\sqrt{3}$. In Rt$\triangle SAC$, $SA = 5\sqrt{3}$, $AC = 5$, then $SC = 10$. In $\triangle SBC$, as $SB^2 = SC^2 + BC^2$, we have SC perpendicular to BC. Then, the surface area of the pyramid is $S = \frac{75}{2} + \frac{25}{2}\sqrt{6} + \frac{25}{2}\sqrt{3}$ cm^2.

11 Draw CD perpendicular to AB with foot of perpendicular D. Since $AC = 20$, $BC = 15$, we have $AB = 25$; then $CD = \frac{20 \times 15}{25} = 12$, $AD = 16$, $BD = 9$. Take the straight line AB as axis, rotate the right-angled triangle ABC for one loop, we get a solid by rotation, which is formed by two cones that share a common base with base radius CD.

$S = 420\pi(\text{cm}^2)$ ⑫ Suppose that the radius of the sphere is r, then the base radius of the cone is r, and its height is $2r$. (a) $\frac{S_{sphere}}{S_{cone\ lateral}} = \frac{4\pi r^2}{\sqrt{5}\pi r^2} = \frac{4}{\sqrt{5}}$

(b) $\frac{S_{sphere}}{S_{cone\ surface}} = \frac{4\pi r^2}{\sqrt{5}\pi r^2 + \pi r^2} = \frac{4}{\sqrt{5}+1} = \sqrt{5} - 1$

6.7 Volumes of 3D shapes

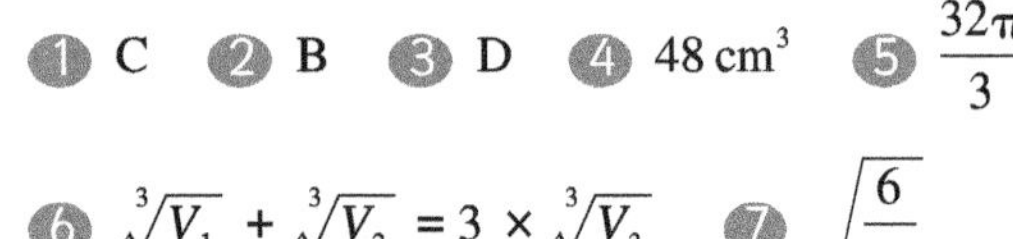

① C ② B ③ D ④ $48\ \text{cm}^3$ ⑤ $\frac{32\pi}{3}$

⑥ $\sqrt[3]{V_1} + \sqrt[3]{V_2} = 3 \times \sqrt[3]{V_3}$ ⑦ $\sqrt{\frac{6}{\pi}}$

⑧ $1 + \frac{\sqrt{2}}{6}$ ⑨ 20 cm ⑩ $\frac{4}{3}\text{cm}^3$ ⑪ Draw an axial cross section of the cone, such that it passes a pair of opposite edges of the cube.

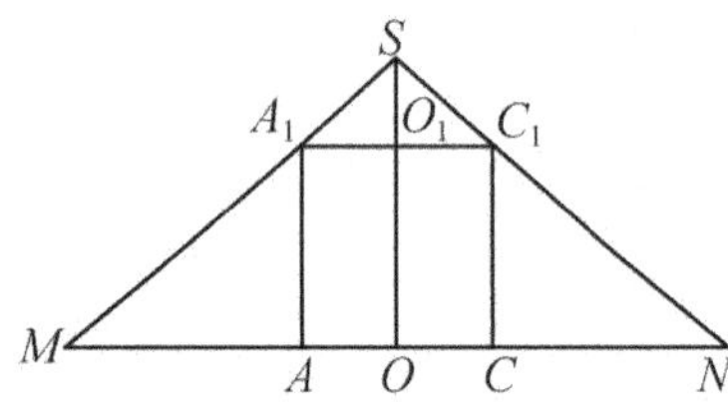

Suppose that the cube has edge length x, then $AC = \sqrt{2}x$, $SM = l$, $\angle SMO = \theta$, then $SO = l\sin\theta$, $MO = l\cos\theta$, Since $\frac{A_1O_1}{MO} = \frac{SO_1}{SO}$, that is $\frac{\frac{1}{2}x}{l\cos\theta} = \frac{l\sin\theta - x}{l\sin\theta}$. Therefore, we have $x = \frac{2l\sin\theta}{2+\tan\theta}$

$\frac{V_{cube}}{V_{cone}} = \frac{x^3}{\frac{1}{3}\cdot\pi\cdot MO^2 \times SO} = \frac{24\tan^2\theta}{\pi(2+\tan\theta)^3}$.

⑫ (a) Suppose that the radius of the cylinder is r. According to the statement in the question, $\frac{r}{2} = \frac{6-x}{6}$, $r = \frac{6-x}{3}$, $S_{cylinder\ lateral} = 2\pi rl = 2\pi \cdot \frac{6-x}{3} \cdot x = -\frac{2\pi}{3}x^2 + 4\pi x\ (0 < x < 6)$ (b) $S_{cylinder\ lateral} = -\frac{2\pi}{3}x^2 + 4\pi x = -\frac{2\pi}{3}(x^2 - 6x) = -\frac{2\pi}{3}(x-3)^2 + 6\pi$. Therefore when $x = 3$, $S_{max} = 6\pi$.

6.8 Measurement involving congruent and similar figures (1)

① B ② B ③ B ④ A ⑤ $18\ \text{cm}^2$ and $27\ \text{cm}^2$ ⑥ $3:7$, $9:49$ ⑦ 4 ⑧ $\sqrt{3}$ ⑨ 4 ⑩ 120 ⑪ $\frac{16}{5}\text{cm}^2$ ⑫ Because the lines are parallel, $\frac{DE}{BC} = \frac{AD}{AB}$, and since $\frac{AD}{BD} = \frac{5}{9}$, $AB = 14\ \text{cm}$. $\frac{AD}{AB} = \frac{5}{14}$, $AD = 5\ \text{cm}$, $BD = 9\ \text{cm}$. As CD is perpendicular to AB, $CD = 12\ \text{cm}$, and $\text{area}_{\triangle ABC} = 84\ \text{cm}^2$, $BC = 15\ \text{cm}$, $AC = 13\ \text{cm}$. $\triangle ABC$ has perimeter $= 42\ \text{cm}$. According to the statements in the question, $\triangle ADE$ is similar to $\triangle ABC$, and the ratio of lengths in the triangles is $\frac{5}{14}$. So the ratio of perimeters = the ratio of line segments $= \frac{5}{14}$, and the perimeter of $\triangle ADE = 15\ \text{cm}$. The ratio of areas is $\frac{5^2}{14^2} = \frac{25}{196}$. So $\text{area}_{\triangle ADE} = 84 \times \frac{25}{196} = \frac{75}{7}\text{cm}^2$. ⑬ (a) $\frac{1}{4}$ (b) No, not changed.

6.9 Measurement involving congruent and similar figures (2)

① D ② D ③ C ④ B ⑤ 48 cm ⑥ $144\ \text{cm}^2$ ⑦ 3 ⑧ $\frac{9-\sqrt{7}}{8}$ ⑨ $\frac{1}{27}$ ⑩ (a) 4 (b) $12 - 6\sqrt{3}$ ⑪ (a) 80 cm, 40 cm (b) $560\ \text{cm}^2$, $140\ \text{cm}^2$ ⑫ 35 cm ⑬ Connect OM, OA. In right-angled triangle SOM, $OM = \sqrt{l^2 - h^2}$. Since the pyramid $SABC$ is a right pyramid, we know that O is the centre of the equilateral triangle ABC. $AB = 2AM = 2OM\tan 60° = 2\sqrt{3} \times \sqrt{l^2 - h^2}$ $\text{area}_{\triangle ABC} = \frac{\sqrt{3}}{4}AB^2 = \frac{\sqrt{3}}{4} \times 4 \times 3(l^2 - h^2) = 3\sqrt{3}(l^2 - h^2)$, $\frac{\text{area}_{\triangle A_1B_1C_1}}{\text{area}_{\triangle ABC}} = \frac{1}{4}$, since the ratio of lengths in triangle $A_1B_1C_1$ to triangle ABC is 1 : 2, so $\text{area}_{\triangle A_1B_1C_1} = \frac{3\sqrt{3}}{4}(l^2 - h^2)$.

Unit test 6

1 C 2 C 3 B 4 A 5 A 6 A 7 C 8 B 9 D 10 $B'C' = 1.5$; $AC = 1.8$ 11 96π 12 33π 13 3.25 14 6 15 40 16 $2\sqrt{5}$ 17 (a) Set the straight line equation as $y = kx + b$, $\begin{cases} -k + b = 0 \\ b = 2 \end{cases}$, then $\begin{cases} k = 2 \\ b = 2 \end{cases}$, so the equation of the straight line is $y = 2x + 2$. (b) $r = \sqrt{2^2 + 1^2} = \sqrt{5}$; the centre of the circle is $(0, 2)$, so the equation of the circle is $x^2 + (y - 2)^2 = 5$. 18 (a) As shown in the diagram

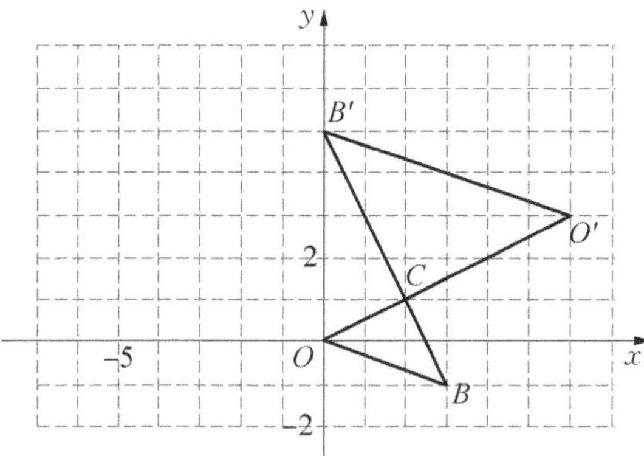

(b) $\text{Area}_{\triangle O'B'C'} = 6 \times 4 - \frac{1}{2} \times 2 \times 4 - \frac{1}{2} \times 4 \times 2 - \frac{1}{2} \times 6 \times 2 = 10$ square units 19 (a) Because $\triangle ABE$ is similar to $\triangle ADB$, $\angle ABD = \angle AEB = 110°$, $\angle D = 180° - \angle ABD - \angle A = 180° - 110° - 40° = 30°$. (b) $\triangle ABE$ is similar to $\triangle ADB$, $\frac{AB}{AD} = \frac{AE}{AB} = \frac{BE}{DB} = \frac{3}{5}$. The ratio of lengths is $\frac{3}{5}$. 20 (a) $\angle ABD = \angle ACB$ (Not the only possible answer.) (b) When $AD : DC = 1 : 2$, $\frac{\text{area}_{\triangle ABD}}{\text{area}_{\triangle CBD}} = \frac{AD}{DC} = \frac{1}{2}$, $\text{area}_{\triangle CBD} = 2\ \text{area}_{\triangle ABD}$, $\text{area}_{\triangle ABC} = \text{area}_{\triangle CBD} + \text{area}_{\triangle ABD} = 3\ \text{area}_{\triangle ABD}$ $\frac{\text{area}_{\triangle ABD}}{\text{area}_{\triangle ACB}} = \frac{1}{3}$. 21 Draw CE perpendicular to AB with the foot of perpendicular E. In $\triangle AEC$, $\angle EAC = 45°$, $AC = 30 \times \frac{40}{60} = 20$, $AE = CE = 20\sin 45° = 10\sqrt{2}$. In $\triangle BEC$, $\angle ECB = 45° + 15° = 60°$, $BE = CE\tan 60° = \sqrt{3}CE = 10\sqrt{6}$. So, $AB = AE + BE = 10\sqrt{2} + 10\sqrt{6} \approx 38.6(\text{km})$.

Chapter 7 Probability

7.1 Systematic listing and the product rule

1 D 2 A 3 A (Hint: there are $3 \times 2 = 6$ ways of arranging the letters in the first column, and for each of these there are 2 possible letters in the second column. There are $= 3 \times 2 \times 2 = 12$ arrangements altogether.) 4 15 5 504 6 12 7 672 8 10 9 162 10 $4 \times 3 \times 5 = 60$ 11 Since $720 = 2^4 \times 3^2 \times 5$, the number of positive divisors $= (4 + 1)(2 + 1)(1 + 1) = 30$. 12 The maximum number of terms is $(n + 1)(m + 1)$, when $f(x) \times g(x)$ is expanded and after collecting the like terms, the smallest number of terms is $n + m + 1$. 13 (a) The number of four-digit palindrome numbers can be obtained by arranging their first two digits of the numbers. There are nine combinations if the first digit is not zero. There are 10 combinations for the second digit. Therefore, there are $9 \times 10 = 90$ combinations for the four-digit palindrome numbers. (b) By studying the data sets, the number of $2n + 1$-digit palindrome numbers is the same as the number of $2n + 2$-digit palindrome numbers. The number of $2n + 2$-digit palindrome numbers follows the pattern of the arrangement of $n + 1$ digits. There are 9 combinations (without 0), and there are 10 combinations for each digit in n-digit. Therefore, the number of $2n + 2$-digit palindrome numbers is 9×10^n, so the number of $2n + 1$-digit palindromes is 9×10^n.

7.2 Probability of mutually exclusive events

1 D 2 C 3 C 4 C 5 $\frac{1}{6}$

6 hitting the target three times 7 0.96

8 0.38 9 Let the events that Chen scores in the test over 90 marks, 80–89 marks, 70–79 marks or 60–69 marks be B, C, D and E, respectively. The four events are mutually exclusive, therefore, the probability of Chen scoring above 80 marks is $P(B) + P(C) = 0.18 + 0.51 = 0.69$. The probability of Chen passing the test, that is, scoring more than 60 marks is $P(B) + P(C) + P(D) + P(E) = 0.18 + 0.51 + 0.15 + 0.09 = 0.93$.

10 Since any two events of the five events are mutually exclusive, based on the addition formula of probability: (a) the probability of hitting 10 or 9 rings is 0.52 (b) the probability of hitting at least 7 rings is $0.24 + 0.28 + 0.19 + 0.16 = 0.87$ (c) the probability of hitting less than 8 rings is $0.16 + 0.13 = 0.29$. 11 (a) $P(A) = \frac{1}{1000}$, $P(B) = \frac{10}{1000} = \frac{1}{100}$, $P(C) = \frac{50}{1000} = \frac{1}{20}$. Then the probabilities of the events A, B and C are $\frac{1}{1000}$, $\frac{1}{100}$ and $\frac{1}{20}$ respectively. (b) A winning lottery ticket may mean winning a grand prize, or a first prize or a second prize. Let the event "A lottery ticket wins a prize" be event M. Then $M = A \cup B \cup C$. Events A, B and C are mutually exclusive, so $P(M) = P(A) + P(B) + P(C) = \frac{1 + 10 + 50}{1000} = \frac{61}{1000}$. So, the probability of a lottery ticket winning a prize is $\frac{61}{1000}$. (c) Let "The probability of a ticket not winning a grand prize nor a first prize" be event N. Then the event N and the event "A ticket wining a grand prize or a first prize" are complementary events, so $P(N) = 1 - P(A \cup B) = 1 - \left(\frac{1}{1000} + \frac{1}{100}\right) = \frac{989}{1000}$. So, the probability of a ticket not winning either a grand prize or a first prize is $\frac{989}{1000}$. 12 The total number of possible outcomes is 25^2. It is possible that they are two white balls, or two red balls or two black balls. Let these events be A, B and C, respectively. They are mutually exclusive. Therefore, the probability of the two balls being the same colour is $P = P(A) + P(B) + P(C) = \frac{3 \times 10}{25 \times 25} + \frac{7 \times 6}{25 \times 25} + \frac{15 \times 9}{25 \times 25} = \frac{207}{625}$.

7.3 Probability of independent and dependent combined events

1 A 2 C 3 C 4 B 5 0.94

6 0.97 7 0.729 8 0.46 9 Let A be the event of one seed for germination from batch M, and B be the event of one seed for germination from the batch N, then $A \cdot B$ is an event in which both seeds can germinate. $A + B$ is an event in which at least one seed can germinate. $A \cdot \overline{B} + \overline{A} \cdot B$ is an event in which exactly one seed can germinate, A and B are independent events, so A and $\overline{B}$, $\overline{A}$ and B, $\overline{A}$ and $\overline{B}$ are mutually independent. (a) $P(A \cdot B) = P(A)P(B) = 0.56$, so the probability of the event that both seeds can germinate is 0.56. (b) $P(A + B) = P(A \cdot B) + P(A \cdot \overline{B}) + P(\overline{A} \cdot B) = P(A)P(B) + P(A)P(\overline{B}) + P(\overline{A})P(B) = 0.56 + 0.8 \times (1 - 0.7) + (1 - 0.8) \times 0.7 = 0.94$ or $P(A + B) = P(A) + P(B) - P(A \cdot B) = 0.8 + 0.7 - 0.56 = 0.94$ or $P(A + B) = 1 - P(\overline{A} \cdot \overline{B}) = 1 - (1 - 0.8)(1 - 0.7) = 0.94$, so the probability of at least one seed germinating is 0.94. (c) $A \cdot \overline{B}$ and $\overline{A} \cdot B$ are mutually exclusive, so $P(A \cdot \overline{B} + \overline{A} \cdot B) = P(A \cdot \overline{B}) + P(\overline{A} \cdot B) = P(A)P(\overline{B}) + P(\overline{A})P(B) = 0.8 \times (1 - 0.7) + (1 - 0.8) \times 0.7 = 0.38$ or $P(A \cdot \overline{B} +$

$\bar{A} \cdot B) = P(A+B) - P(A \cdot B) = 0.94 - 0.56 = 0.38$. Hence, the probability that exactly one seed can germinate is 0.38. **10** The probabilities that each person cannot decipher the password are $\frac{4}{5}$, $\frac{3}{4}$ and $\frac{2}{3}$ respectively. These three events are independent, so the probability that the password cannot be deciphered is $\frac{4}{5} \times \frac{3}{4} \times \frac{2}{3} = \frac{2}{5}$. Therefore the probability that the password can be deciphered is $1 - \frac{2}{5} = \frac{3}{5}$. **11** The probability that the first number obtained is even is $\frac{1}{2}$, the second is odd is $\frac{1}{2}$, and the third number greater than 4 is $\frac{1}{3}$. Then the probability is $\frac{1}{2} \times \frac{1}{2} \times \frac{1}{3} = \frac{1}{12}$. (The tree diagram is omitted.) **12** Let the probability that event A occurs in one experiment be P. Then the probability that A does not occur in one experiment is $1 - P$. Therefore, the probability that event A does not occur in the 4 individual experiments is $(1-P)^4$. The probability of A occurring the minimum number of times is $1 - (1-P)^4 = \frac{65}{81}$, which gives $P = \frac{1}{3}$. Therefore, the probability that event A occurs in an experiment is $\frac{1}{3}$. **13** Denote the components A, B, and C working normally as events A, B, and C respectively. From the given information, $P(A) = 0.80$ and $P(B) = P(C) = 0.9$. Since events A, B, and C are mutually independent, the probability of the system N_1 working normally is $P_1 = P(A \cdot B \cdot C) = P(A)P(B)P(C) = 0.80 \times 0.90 \times 0.90 = 0.648$. The probability of the system N_2 working normally is $P_2 = P(A) \cdot [1 - P(\bar{B} \cdot \bar{C})] = P(A) \cdot [1 - P(\bar{B}) \cdot P(\bar{C})] = 0.80 \times [1 - (1 - 0.90)(1 - 0.90)] = 0.792$.

7.4 Theoretical probability and experimental probability

1 D **2** A **3** C **4** D **5** $\frac{1}{4}$ **6** 0.72 **7** 0.76 **8** (a) 0.49 0.54 0.50 0.50 (b) 0.5 **9** 0.53 **10** (a) 0.2 (b) 0.3 (c) The probability of the customer purchasing products A and B at the same time can be estimated as $\frac{200}{1000} = 0.2$, the probability of the customer purchasing products A and C at the same time can be estimated as $\frac{100 + 200 + 300}{1000} = 0.6$, and the probability of the customer purchasing products A and D at the same time can be estimated as $\frac{100}{1000} = 0.1$. Therefore, if the customer purchased product A, then they are most likely to purchase product C. **11** (a) 0.852 (b) 25 560 freshwater fish (c) About 5869 **12** (a) The frequencies of good-quality balls in the table are 0.900, 0.920, 0.970, 0.940, 0.954, 0.951 respectively. (b) The total number of balls in the samples is 3850. Of these, the total good quality balls is 3657. So the probability of getting a good quality ball is $3657 \div 3850 = 0.950$ (to three decimal places).

7.5 Conditional probabilities

1 C **2** D **3** D **4** $\frac{12}{19}$ **5** $\frac{9}{17}$ **6** $\frac{9}{70}$ **7** $\frac{10}{19}$ **8** $\frac{2}{3}$ **9** (a) The total number of students surveyed is 28, and the total number of students with two siblings is 3. So the probability that a student in the class has 2 siblings is $P(2) = \frac{3}{28}$. (b) The probability that a female student in the class has 2 siblings is $\frac{2}{15}$. (c) The probability that a male student in the class has no siblings is

$\frac{9}{13}$. **10** Let A = "It is a rainy day in city M" and B = "It is a rainy day in city N". Then from the question we get $P(A) = 0.20$, $P(B) = 0.18$ and $P(A \cap B) = 0.12$ (a) The probability that it rained in M given that it rained in N is $P(A \mid B) = \frac{P(A \cap B)}{P(B)} = \frac{0.12}{0.18} = 0.67$. (b) The probability that it rained in N given that it rained in M is $P(B \mid A) = \frac{P(A \cap B)}{P(A)} = \frac{0.12}{0.20} = 0.60$. **11** Let B be the grade I piece and A be the qualified piece. (a) Since 70 items are classified as grade I out of 100 items of the product, $P(B) = \frac{70}{100} = 0.7$.

(b) $P(B \mid A) = \frac{P(B)}{P(A)} = \frac{\frac{70}{100}}{\frac{95}{100}} = \frac{14}{19}$. **12** (a) Let "The sum of the numbers is 7" be event A and "One of the dice showing 2" be event B. Then the possible outcomes of event A are (1, 6), (2, 5), (3, 4), (4, 3), (5, 2), (6, 1), so there are 6 possible outcomes. Event $B \mid A$ has 2 possible outcomes, so $P(B \mid A) = \frac{2}{6} = \frac{1}{3}$. Therefore, when the sum of the numbers is 7, the probability that one of the dice shows 2 is $\frac{1}{3}$. (b) Let "When the two dice show different numbers" be event C. Then event C has 30 possible outcomes. Let "one of the dice shows 4" be event D. Then the possible outcomes of D given C are (1, 4), (2, 4), (3, 4), (5, 4), (6, 4), (4, 1), (4, 2), (4, 3), (4, 5), (4, 6). There are 10 possibilities in total. Therefore, $P(D \mid C) = \frac{10}{30} = \frac{1}{3}$. So, when the two dice show different numbers, the probability of one of the dice showing 4 is $\frac{1}{3}$.

Unit test 7

1 A **2** C **3** C **4** D **5** $\frac{1}{3}$

6 $\frac{1}{3}$ **7** $\frac{1}{2}$ **8** $\frac{2}{5}$ **9** 0.3 **10** ②

11 ①②③ **12** 0.5 **13** 0.35 **14** (a) $\frac{1}{12}$

(b) $\frac{1}{2}$ (c) $\frac{3}{5}$ **15** Let the event "fail in the mathematics test" be event A and the event "fail in the English test" be B. Then $P(A) = 0.15$, $P(B) = 0.05$ and $P(AB) = 0.03$. Therefore,

(a) $P(B \mid A) = \frac{P(AB)}{P(A)} = \frac{0.03}{0.15} = 0.2$ (b) $P(A \mid B) = \frac{P(AB)}{P(B)} = \frac{0.03}{0.05} = 0.6$. **16** (a) The sample size is 30, and the number of days without rain is 26. The probability is estimated using relative frequency. The probability of no rain on a randomly selected day in this city is $\frac{26}{30} = \frac{13}{15}$.

(b) In April, a day is sunny, then there are 16 pairs of days to consider (a sunny day and the day following this) of which there are 14 pairs of days when it does not rain on the day following a sunny day. So the relative frequency of not raining on the day following a sunny day is $\frac{7}{8}$. So the probability of not raining during the games is $\frac{7}{8}$. **17** (a) Let "score 9 or 10 points from one shot" be event A. Events A_9 and A_{10} are mutually exclusive, so $P(A) = P(A_9) + P(A_{10}) = 0.28 + 0.32 = 0.60$.

(b) Let the event of "score 8 points or more from one shot" be event B. Then $\bar{B}$ represents the event "score less than 8 points from one shot". $P(B) = P(A_8) + P(A_9) + P(A_{10}) = 0.18 + 0.28 + 0.32 = 0.78$. So $P(\bar{B}) = 1 - P(B) = 1 - 0.78 = 0.22$. So the probability of scoring less than 8 points from one shot is 0.22.

Chapter 8 Statistics

8.1 Sample and population

1 D 2 D 3 C 4 B 5 C
6 population 7 random sampling 8 all the students, the 250 students selected.
9 (a) population: all the adults across a country, sample: 868 adults selected. (b) population: all the families across this country, sample: 1220 families selected in this city 10 Select stratified sampling, the total number of students is 1220, the number of students to be sampled is 183, and 36, 36, 38, 37, 36 students should be selected from year groups 7 to 11 respectively.
11 Selecting the whole population takes a lot of time and effort, and sometimes it is impractical to investigate the whole population.

8.2 Tables and line graphs for time series data

1 C 2 C 3 A 4 D 5 (a) 2.7 (b) 2015 to 2016 (c) 2016 to 2017 6 236, 292, 815, 393 7 (a) Graph omitted.
(b) (i) the table (ii) The line graph because it shows the trend at a glance. 8 (a) The growth rate of the GDP was 2.30%, 2.11%, 1.65%, 1.67%, −9.27%, 7.43%, respectively from 2016 to 2021 (b) Graph omitted. (c) 2.85%; mean annual growth rate = 0.98% (d) (i) 2020 to 2021 had the largest increase. (ii) 2019 to 2020; the COVID-19 pandemic may be one of the causes. 9 (a) Graph omitted. (b) (i) 2009; 9.9% (ii) 3.9% in 2018 (c) The US unemployment rate steadily decreased from year 2009.

8.3 Statistical graphs for grouped discrete data

1 C 2 D 3 B 4 D 5 A
6 40, 28, 35, 48 7 35 8 64, 96
9 223 10 (a) a = 24. b = 40 (b) Histogram omitted. (c) Most cars produced by the factory are between 3.0 m and 3.2 m in length.
11 (a) Students arrived 0 to 2 minutes earlier than the starting time of the event. (b) (i) Answer omitted. (ii) median approximately 2.3 minutes; IQR approximately 3.4 minutes

8.4 Graphical representation of the distributions of statistical data (1)

1 D 2 C 3 C 4 B 5 A = 69, B = 73.5, C = 78, D = 82.5, E = 89, IQR = 9
6 A − matches the second box plot, B − matches the third box plot, C − matches the first box plot.
7 (a) 300 mm (b) median (c) 90 (d) 50%
8 (a) The lower quartile is 54 minutes. (b) The shortest time was 30 minutes, the lower quartile is 54 minutes, the median was 3 hours 36 minutes, the upper quartile was 4 hours 42 minutes and the longest time taken was 5 hours 12 minutes.
9 (a) Team A: the lightest mass is 62 kg, the lower quartile is 64 kg, the median is 67 kg, the upper quartile is 70.5 kg and the greatest mass is 72 kg. Team B: the lightest mass is 64 kg, the lower quartile is 64.5 kg, the median is 65 kg, the upper quartile is 67.5 kg and the greatest value is 70 kg. (b) Team A: the range is 10 kg, the interquartile range is 6.5 kg, the median is 67 kg. Team B: the range is 6 kg, the interquartile range is 3 kg, the median is 65 kg. Players in team B have a more concentrated mass distribution, i. e. less variation in their masses. Players in team A have a larger average (median) mass.

8.5 Graphical representation of the distributions of statistical data (2)

1 A 2 C 3 B 4 D 5 A
6 $15 \leqslant x < 20$ 7 median = 4; IQR = 4
8 $x = 10$, $y = 13$ 9 The values of a, b, c and d are 3, 6, 8, 8, respectively 10 4, 28
11 (a) 40 cm (b) 30 cm, 60 cm (c) 35 (d) 80, 35 (e) 45.5 cm

8.6 Apply statistics to describe a population

1 A 2 D 3 D 4 A 5 central tendency (mean or median); spread (range or interquartile range) 6 80, 78.75 7 possible

8 not true 9 (a) size 7 (b) 225; $\frac{15}{100} \times 1500 = 225$ 10 (a) Robert: the median score is 3 and interquartile range is 5.1; Nala: the median score is 8 and interquartile range is 4.3 (b) boxplot (c) Robert; he gave 6 students a score greater than 9, whereas Nala gave 32 students a score greater than 9.

8.7 Scatter graphs of bivariate data (1)

1 C 2 A 3 C 4 B 5 D 6 independent, dependent 7 increases 8 decreases 9 (a) Answer omitted. (b) Positive (c) The taller a student is, the greater their mass. (d) A strong positive correlation 10 (a) Answer omitted. (b) 396 km, 28 litres (c) Answer omitted. (d) Approximately 33 litres

8.8 Scatter graphs of bivariate data (2)

1 C 2 C 3 D 4 a weak positive correlation 5 no correlation 6 a strong positive correlation 7 a strong negative correlation 8 (a) 0.664. It means that for each increase of 1 unit in the IQ values, the estimated science aptitude test scores increase by 0.664. (b) 2.39. No, it is not suitable. (c) Yes, they are. No, there is no causation between them. 9 (a) Answer omitted. (b) 64.25, 65.125 (c)(d) Answer omitted. (e) Approximately 51 (f) No; there is no data recorded for scores greater than 82 in either subject, so you cannot be sure that the trend will continue.

Unit test 8

1 B 2 A 3 B 4 D 5 D 6 D 7 mean, median, mode 8 0.5, 6.5, 6 9 (a) 81 (b) 75.1 (c) Answer omitted (d) 71–80 10 (a) Stratified sampling method, it can obtain a representative sample from a population. (b) A: 19, B: 26, C: 39, D: 55, E: 61 11 (a) $95 \leqslant s < 110$ (b) Answer omitted (c) 20 12 (a) (4, 30) (b) Answer omitted (c) The more time a person spends in a gym, the lower the body mass index they attain. (d) Approximately 20; fairly accurate because it is obtained by interpolation (6 hours is within the range of the data collected).

End of year test

1 B 2 C 3 D 4 D 5 A 6 D 7 D 8 B 9 C 10 A 11 A 12 $y = -3x - 7$ 13 $k > 8$ or $k < -4$ 14 $\frac{4}{5}$ 15 $2a + \sqrt{2}b$ 16 9 : 16 17 $m < \frac{3}{2}$ 18 $\frac{1}{3}$ 19 $f(g(x)) = \sqrt{1 - \cos x}$ 20 $\frac{18}{5}$ cm, $\frac{12}{5}$ cm 21 9π 22 $\frac{\sqrt{2}}{2}$ 23 14 24 $\frac{8}{25}$ 25 4

26 When $x = 1$, $x^3 - x + 1$ is $-1 < 0$, and when $x = 2$, $x^3 - x + 1$ is $5 > 0$. Therefore, the equation $x^3 - x + 1 = 2$ has a solution between $x = 1$ and $x = 2$. 27 (a) The 5th term is 19 and the 8th term is 10, so $a_1 + 4d = 19$, $a_1 + 7d = 10$, so $a_1 = 31$ and $d = -3$. So the 1st term is 31 and the common difference is -3. (b) The 4th item is 19 and the 10th item is 4, so $b_1 + 3d = 10$, $b_1 + 9d = 4$, so $b_1 = 13$ and $d = -1$ and $b_{14} = 13 - (14 - 1) = 13 - 13 = 0$. 28 In the right-angled triangle $\triangle ABD$, $\angle ABD = 30°$, and $AD = 6$ m $\frac{AD}{AB} = \sin 30° = \frac{1}{2}$, so $AB = 2AD = 12$ m. In the right-angled triangle ACD, $\angle ACD = 16°$ and $AD = 6$ m. $\frac{AD}{AC} = \sin\angle ACD$, so $\frac{6}{AC} = \sin 16°$, which gives $AC = \frac{6}{\sin 16°} = 21.767\cdots$. Then $AC - AB = 21.767\cdots - 12 = 9.77$ m (3 s. f.), so the new escalator needs to be 9.77 m longer than the existing one. 29 (a) $50 \leqslant w < 55$ (b) Histogram omitted. (c) 9 30 (a) Put $A(-1, 0)$, $B = (2, 0)$,

$C(0, 2)$ into $y = ax^2 + bx + c$, then $\begin{cases} 0 = a - b + c \\ 0 = 4a + 2b + c \\ c = 2 \end{cases}$, which gives $a = -1$, $b = 1$ and $c = 2$, so the equation of the quadratic function is $y = -x^2 + x + 2$. (b) The line of symmetry goes through the midpoint of A and B. $\frac{-1 + 2}{2} = \frac{1}{2}$ so the equation of the line of symmetry is $x = \frac{1}{2}$. Substituting $x = \frac{1}{2}$ into $y = -x^2 + x + 2$ gives $y = -\left(\frac{1}{2}\right)^2 + \frac{1}{2} + 2 = \frac{9}{4}$ so the coordinates of the vertex are $\left(\frac{1}{2}, \frac{9}{4}\right)$.

31 (a) Put $x = 0$ into $y = kx + 4$. Then $y = 4$ and $OB = 4$. Since $\angle BAO = 45°$ and $\angle AOB = 90°$, $OA = OB = 4$. $A(-4, 0)$. Put $x = -4$, $y = 0$ into $y = kx + 4$. Then $k = 1$.

(b)

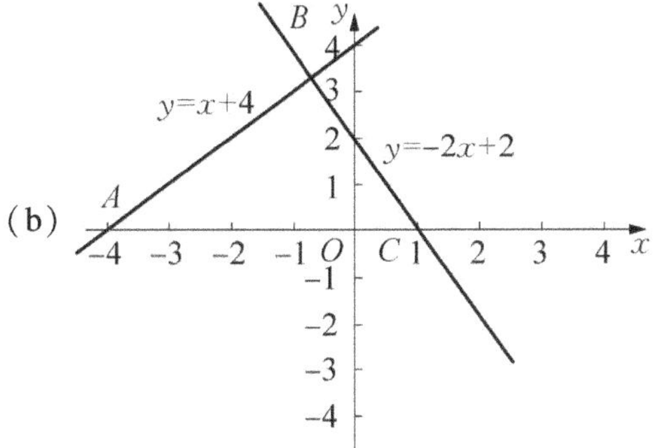

(c) $y = -2x + 2$ intersects the x-axis at point C and $y = x + 4$ intersects at point P. Substituting $y = 0$ into $y = -2x + 2$, gives $x = 1$. So $C(1, 0)$, and $AC = 5$. To find where the two lines intersect, solve simultaneously: $\begin{cases} y = x + 4 \\ y = -2x + 2 \end{cases}$. This gives $\begin{cases} x = -\frac{2}{3} \\ y = \frac{10}{3} \end{cases}$ $\text{area}_{\triangle PAC} = \frac{1}{2} \times 5 \times \frac{10}{3} = \frac{25}{3}$ square units.

32 (a) $\triangle ACD$ has the largest area when AD is perpendicular to AC. Therefore angle $ADC = 90°$. Let $AD = x$; $AD = AB = x$, so $AC = 2AB = 2x$, and $\sin C = \frac{AD}{AC} = \frac{x}{2x} = \frac{1}{2}$. (b) $\angle A = 60$; reason: when CD is the tangent of circle A, then AD is perpendicular to DC. Since $AB = BC$ and $AD = AB$, we get $AD = \frac{1}{2}AC$, so $\cos A = \frac{x}{2x} = \frac{1}{2}$; hence $\angle A = 60°$.

33 (a) We need to find the probability that the amount of compensation is either £3000 or £4000. Let A denote the event "the amount of compensation is £3000", and B denote the event "the amount of compensation is £4000". Using relative frequency to estimate the probability of each event, $P(A) = \frac{150}{1000} = 0.15$ and $P(B) = \frac{120}{1000} = 0.12$. So the required probability is $P(A) + P(B) = 0.15 + 0.12 = 0.27$. (b) The number of vehicles in the sample whose owners are new drivers is $0.1 \times 1000 = 100$, and the number of the vehicles whose owners are new drivers and received £4000 compensation is $0.2 \times 120 = 24$. So, given that an insured car owner in the sample is a new driver, the probability that they received £4000 compensation $= \frac{24}{100} = 0.24$.